GENERATIVE AI FOR BEGINNERS

ESCAPE THE LEARNING CURVE!

A 7-Day Guide to Learn AI, Deep Learning, and ChatGPT — Master Artificial Intelligence and Build Creative Projects Even If You're Not Tech-Savvy

JOHN COOK

TOP TIER PRESS

CONTENTS

Introduction ix
Unlock Your AI in Action Guide xv

1. ESCAPE THE OVERWHELM 1
Understanding AI in Plain English
The Great AI Intimidation 2
AI Simplified: Three Definitions That Actually Make Sense 2
AI is Already Your Invisible Assistant 5
Why Recognition Matters: Building Confidence in an AI World 8
The Confidence Foundation 9

2. MEET YOUR AI TOOLKIT 11
ChatGPT, Midjourney & More
The Big Three: Your AI Starting Lineup 12
The Supporting Cast: Specialized AI Tools Worth Knowing 15
Free vs. Premium: Making Smart Investment Decisions 18
Quick Setup: Your 30-Minute AI Toolkit Installation 19
Your Next Steps: From Setup to Mastery 22

3. DAY 1 25
Your First Conversation with ChatGPT
The Art of Talking to AI: Understanding Prompts 26
Your First Real Conversation: A Step-by-Step Walkthrough 27
Hands-On Project: Creating Your First AI-Assisted Short Story 29
The Q&A Deep Dive: Exploring What You've Always Wondered 31
Key Lesson: AI Isn't Magic, It's a Tool 32
What You've Accomplished Today 34

4. DAY 2 37
Productivity Boost with AI
The Productivity Revolution You Didn't See Coming 37
Writing That Works: Beyond Generic Templates 38
The Mini-Project: AI Organizes Your Schedule 40
From Overwhelm to Control: The Mental Shift 42
Real-World Applications: Where the Magic Happens 43
Advanced Productivity Techniques 45
Building Your AI Productivity System 46
Overcoming Common Obstacles 47
Tomorrow's Foundation 48

5. DAY 3 49
Creativity Unleashed
The Creative Revolution You Can Join Today 49
Writing with AI: From Blank Page to Published Post 50
Visual Creativity: Bringing Ideas to Life 52
Mini-Project: Crafting Your First AI-Assisted Blog Post 54
Building Creative Confidence Through Experimentation 57
Your Creative Toolkit Expanding 58

6. DAY 4 59
Deep Learning Without the Math
The Brain That Isn't Really a Brain 59
The Magic Behind the Curtain 60
Real-World Deep Learning in Action 61
Your First AI Art Adventure 63
The Deeper Understanding 66
Beyond the Hype: Real Limitations 67
Looking Forward: Your Next Steps 68

7. DAY 5 71
Build Your First AI Project
Why Multi-Tool Projects Change Everything 71
Three Project Blueprints to Get You Started 72
Your First Mini-Project: The 90-Minute Challenge 78
Measuring Your Success 81
Looking Beyond Your First Project 81
Your AI Project Journey Starts Now 83

8. DAY 6 85
Smarter, Safer AI Use
Understanding AI's Fundamental Limitations 86
Fact-Checking and Verification Strategies 89
Avoiding Dangerous Overreliance 91
Ethical Guidelines for Different User Groups 93
Building Sustainable AI Habits 97

9. DAY 7 99
Graduation Day – Your Final Challenge
Welcome to Your AI Graduation 99
Your Journey: From Zero to AI-Confident 100
Your Final Challenge: Choose Your Path 101
Pathway 1: The Productivity Powerhouse 102
Pathway 2: The Creative Catalyst 104
Pathway 3: The Learning Accelerator 105
Executing Your Chosen Project 107
The Deeper Transformation 108
Your Ongoing AI Journey 109
The Bigger Picture: Your Role in the AI Future 111
Completing Your Final Project 112

10. AI AT WORK 115
Careers & Business Opportunities
How Industries Are Adopting AI 116
Future-Proofing Your Career 119
How to Stay Ahead as AI Evolves 122

11. AI FOR PERSONAL GROWTH & EVERYDAY LIFE 127
The Rise of AI as Your Personal Growth Partner 128
AI as Your Personal Assistant: Beyond Scheduling and Reminders 128
AI as Your Tutor: Personalized Learning Revolution 129
AI as Your Coach: Motivation, Accountability, and Growth 131
Using AI for Hobbies and Creative Exploration 132
Self-Improvement Through AI-Guided Reflection 133
Designing Your AI-Enhanced Lifestyle: A Practical Framework 134
Building Your AI-Enhanced Life: A Step-by-Step Mini-Project 138
The Path Forward 140

12. FROM BEGINNER TO CONFIDENT CREATOR 143
The Hidden Traps That Trip Up New AI Users 144
Building Consistency Without Burning Out 146
Your Curated Learning Path Forward 147
Making AI Work for Your Specific Situation 151
Your Next Steps 152

Guide a Curious Mind Into the World of AI 155

CONCLUSION 157
AI as Your Creative and Professional Partner 158
Your Invitation to Keep Exploring and Creating 159
The Confidence to Shape Your AI-Enhanced Future 160

AI

FOR

EGIN

INTRODUCTION

If you've picked up this book, chances are you've found yourself in one of these situations: You're at a dinner party, and everyone's talking about ChatGPT and how it's "changing everything," while you nod along wondering what the fuss is about. Your teenager shows you an AI-generated image that looks impossibly real, leaving you both amazed and slightly unsettled. Or maybe your boss mentioned something about "leveraging AI for productivity," and you panicked, wondering if you're about to be left behind in the digital dust.

Take a deep breath. You're not alone, and you're definitely not too late to the party.

The Great AI Myth-Busting Session

Let's start by demolishing some of the biggest myths that might be keeping you from diving into the world of artificial intelligence.

Myth #1: "AI is Only for Tech Experts"

This is perhaps the most damaging myth of all. Yes, *creating* AI systems requires advanced degrees and years of coding experience. But *using* AI? That's as simple as having a conversation or uploading a photo. Think about it this way: you don't need to understand how a car engine works to drive to the grocery store. Similarly, you don't need a computer science degree to ask ChatGPT to help you write a professional email or use Midjourney to create artwork for your small business.

The AI tools available today are designed for regular people. Companies like OpenAI, Google, and others have spent millions of dollars making sure their interfaces are user-friendly. They *want* you to use their products, which means they've made them accessible to everyone.

Myth #2: "It's Too Technical and Complicated"

Here's the truth: most AI interactions happen through simple text conversations. If you can send a text message or write an email, you can use AI. The days of needing to understand complex algorithms or write code are over for everyday users. Modern AI tools are designed with intuitive interfaces that respond to plain English.

You might be thinking, "But I've seen those technical articles with all the jargon!" Yes, those exist, but they're written for developers and researchers, not for people who simply want to use AI to make their lives easier. It's like the difference between reading a car repair manual and learning how to use GPS navigation.

Myth #3: "AI Will Replace Me"

This fear runs deep, and it's understandable. But here's what the data actually shows: AI is much better at *augmenting* human capabilities than replacing them entirely. The people who will thrive in the AI era are those who learn to work alongside AI, not those who ignore it or those who expect it to do everything for them.

Think of AI as the ultimate assistant – one that never gets tired, never judges your questions, and is available 24/7. It can help you brainstorm ideas, draft documents, solve problems, and even learn new skills. But it still needs your creativity, judgment, and human touch to create something truly valuable.

What This Book Will Actually Do for You

Forget everything you think you know about learning new technology. This isn't a textbook, and you won't be memorizing technical specifications or writing a single line of code.

Your 7-Day Transformation

This book is built around a simple premise: **you can confidently become AI-literate in just seven days**. Not "expert level" – we're talking about practical, everyday confidence that will make you feel like you're finally catching up with the modern world.

Here's what each day will bring you:

Day 1: You'll have your first real conversation with AI and discover it's nothing like the scary robots from movies. By evening, you'll understand what AI actually is and isn't.

Day 2: You'll learn the art of prompt engineering – essentially, how to ask AI the right questions to get the answers you want. This single skill will unlock 80% of AI's practical value for you.

Day 3: We'll focus on using AI for writing tasks, from emails to creative projects. You'll discover how AI can help you find the right words when you're stuck.

Day 4: Visual AI takes center stage. You'll create your first AI-generated images and learn how these tools can enhance your personal and professional projects.

Day 5: Time to get practical with productivity. You'll discover specific ways AI can streamline your work, organize your life, and save you hours each week.

Day 6: We'll explore AI's role as a learning companion. Whether you want to understand a complex topic or develop a new skill, you'll learn how to turn AI into your personal tutor.

Day 7: Integration day. You'll create a personalized AI workflow that fits your specific needs and goals.

The "Quick Win" Philosophy

Every chapter in this book is designed around immediate usability. You won't just read about concepts – you'll actually use AI tools and see results within minutes. Each day builds on the previous one, but every single day also delivers standalone value that you can apply immediately.

This approach is intentional. Adult learners need to see progress quickly, especially when tackling something that initially feels intimidating. By the end of Day 1, you'll have already accomplished something tangible with AI. By Day 3, you'll be using it for real tasks. By the end of the week, AI will feel like a natural part of your toolkit.

Your Promise: Real Confidence in Real Life

Here's my commitment to you: by the end of this book, you'll be the person others come to with AI questions. Not because you've become a technical expert, but because you've developed practical fluency with the tools that matter.

What "AI Confidence" Actually Looks Like

You'll know you've succeeded when:

- You can help a colleague craft a difficult email using AI assistance - You're comfortable experimenting with new AI tools as they emerge - You can explain to your friends what AI can and can't do (without the hype or the fear) - You have a handful of go-to AI workflows that make your daily life easier - You feel excited, not anxious, when you hear about new AI developments

Beyond the Hype and Horror Stories

The media loves extreme stories about AI – either it's going to solve all our problems or it's going to end civilization as we know it. The reality is much more mundane and much more useful. AI is a powerful tool that, like any tool, produces results based on how skillfully you use it.

This book will help you cut through the noise and focus on what actually matters: practical applications that improve your life starting today. We'll acknowledge AI's limitations honestly, celebrate its genuine capabilities, and always keep the focus on real-world usefulness.

Ready to Begin?

If you've made it this far, you're already demonstrating the curiosity and openness that will make you successful with AI. You don't need to be the most tech-savvy person in the room. You don't need to understand machine learning algorithms. You just need to be willing to try something new and give yourself permission to learn at your own pace.

The AI revolution isn't coming – it's already here. But unlike previous technological shifts, this one has been designed to include everyone. The question isn't whether you're smart enough or technical enough to participate. The question is simply: are you ready to discover what AI can do for you?

Turn the page, and let's find out together.

Tomorrow, we'll start with Day 1: your first real conversation with AI. Bring your curiosity and leave your preconceptions behind. You might be surprised by how natural it feels.

UNLOCK YOUR AI IN ACTION GUIDE

Ready to move from curiosity to confident creation with AI?

You've already explored the foundations of generative and agentic intelligence. Now it's time to put them into motion. ***AI in Action*** gives you clear, step-by-step strategies and real examples to help you innovate faster, simplify your workflow, and bring your ideas to life with confidence and clarity.

What's included in this guide::

- Accelerate creativity and focus
- Automate tasks with precision
- Learn smarter every single day
- Build confidence using real tools

👉 **Access Your AI in Action Guide Now**

Visit: **https://content.toptierpress.com/unlock-ai-action-guide**

Or scan the QR code below to get instant access!

[1]

ESCAPE THE OVERWHELM

UNDERSTANDING AI IN PLAIN ENGLISH

The notification popped up on Sarah's phone: "Netflix has new recommendations for you based on your viewing history." She tapped it, found the perfect show for her evening wind-down, and settled in without giving it another thought. Meanwhile, her husband, Mark, was asking Siri about tomorrow's weather while their teenage daughter, Emma, scrolled through an Instagram feed perfectly curated to her interests, photos of indie bands, vintage fashion, and local coffee shops she'd never explicitly told the app she cared about.

Three people, three different interactions, all powered by artificial intelligence. Yet none of them would have described their evening as "using AI." This disconnect between our daily reality and our perception of AI is at the heart of why so many people feel overwhelmed when they hear terms like "machine learning," "neural networks," or "generative AI." We're not behind on technology, we're already living with it. We just don't recognize it yet.

THE GREAT AI INTIMIDATION

Let's address the elephant in the room. When most people hear "artificial intelligence," their minds conjure images of robots taking over the world, complex computer code scrolling down black screens, or impossibly smart machines that only PhD scientists can understand. Hollywood hasn't helped, from HAL 9000 in *2001: A Space Odyssey* to the more recent fears about AI dominating humanity, popular culture has painted AI as either magical or menacing.

The reality is far more mundane and far more useful. AI isn't some distant, incomprehensible technology that requires years of study to understand. It's a collection of tools and techniques that help computers recognize patterns, make predictions, and solve problems in ways that feel surprisingly human-like. The "intelligence" in artificial intelligence isn't about consciousness or sentience, it's about usefulness.

Think of AI as a very sophisticated pattern-matching system. Just like you can recognize your friend's voice on the phone without seeing them, or predict that someone holding dark clouds usually means rain is coming; AI systems learn to spot patterns in data and make educated guesses about what comes next.

AI SIMPLIFIED: THREE DEFINITIONS THAT ACTUALLY MAKE SENSE

Artificial Intelligence: The Umbrella Term

Artificial Intelligence (AI) is any computer system that can perform tasks we typically associate with human thinking, recognizing speech, understanding images, making decisions, or solving problems. It's the broad umbrella under which everything else falls.

Here's the key insight: AI doesn't mean the computer is actually thinking like a human. It means the computer can produce results that look like they came from human thinking. When your email automatically sorts spam from legitimate messages, that's AI. The computer isn't "understanding" your email the way you do. It's analyzing hundreds of signals (sender reputation, word patterns, link structures) to make a classification that feels intelligent to you.

The goal of AI has always been practical: **make computers more useful by making them better at tasks that humans excel at naturally**. We're good at recognizing faces, understanding speech, and making sense of messy, incomplete information. Traditional computers, for all their computational power, have historically been terrible at these things. AI bridges that gap.

Machine Learning: AI That Improves With Experience

Machine Learning is a subset of AI where computers learn to make better predictions by analyzing examples, rather than following pre-programmed rules. Instead of a human programmer writing out every possible scenario and response, the system learns patterns from data.

Imagine teaching a child to recognize different dog breeds. You could try to write out detailed rules: "Golden Retrievers have long, golden fur and floppy ears, while Chihuahuas are small with pointy ears..." But that approach breaks down quickly. There are hundreds of breeds, lots of mixed breeds, and endless variations within each type.

Machine learning takes a different approach. You show the system thousands of photos labeled with the correct breed, and it gradually learns to associate visual patterns with labels. After enough examples, it can identify breeds in new photos it has never seen before. The system doesn't know "rules" about dog breeds in the way humans do. It knows statistical relationships between visual features and breed labels.

This is exactly how your photo app organizes your pictures by recognizing faces, or how streaming services predict what you might want to watch next. The systems have learned from millions of examples of what people with similar viewing histories enjoyed.

Generative AI: The Creative Assistant

Generative AI is the newest addition to mainstream AI tools, and it's probably what prompted you to pick up this book. This type of AI doesn't just classify or predict—it creates. Generative AI can write stories, compose music, generate images, or even produce computer code.

The breakthrough that made generative AI possible was teaching computers to understand and manipulate language, images, and other creative mediums at a level that feels remarkably human-like. Tools like ChatGPT can write emails, essays, and poetry. DALL-E can create original artwork from text descriptions. GitHub Copilot can help programmers write software.

What makes generative AI fascinating is that it doesn't just follow templates or combine existing pieces. It appears to understand context, style, and intent in ways that produce genuinely novel outputs. Ask it to write a haiku about pizza, and it doesn't retrieve a pre-written pizza haiku from a database. It generates original lines that follow haiku structure while capturing something meaningful about pizza.

The word "generative" is key here. These systems generate new content based on patterns they've learned from vast amounts of existing content. They're like extremely well-read, creative assistants who can adapt their writing style, artistic approach, or problem-solving method to match what you need.

AI IS ALREADY YOUR INVISIBLE ASSISTANT

The most surprising thing about AI isn't how futuristic it seems. It's how seamlessly it has woven itself into the fabric of everyday life. You've been using AI for years, probably multiple times every day, without thinking about it as AI at all.

Your Personal Entertainment Curator

Every time you open Netflix, YouTube, or Spotify, AI systems are working behind the scenes to personalize your experience. Netflix's recommendation algorithm analyzes not just what you've watched, but how you watched it. Did you binge the entire season in one sitting? Skip the intro? Stop watching halfway through episode three? All of these signals feed into a complex model that tries to predict what will keep you engaged.

The algorithm considers hundreds of factors: the time of day you usually watch, the genres you prefer on weekdays versus weekends, how your viewing habits compare to similar users, and even subtle factors like whether you tend to watch comedies when you're stressed. It's not just matching you with popular content. It's trying to understand your personal taste and mood patterns.

YouTube takes this even further. Its AI doesn't just recommend videos based on what you've watched. It analyzes how long you watched, whether you liked or shared content, and even what you searched for but didn't click on. The goal is to create a personalized video feed that keeps you engaged, and the system gets better at this task with every interaction.

The Search Engine That Reads Your Mind

Google's search functionality is perhaps the most sophisticated AI system most people interact with daily. When you type "good Italian restaurant nearby," Google's AI needs to understand that you're not looking for articles about Italian cuisine or the history of Italian immigration. You want restaurant recommendations in your current location.

The search AI processes natural language, understands context and intent, considers your location and search history, and cross-references millions of web pages, reviews, and business listings to provide relevant results in milliseconds. It can even handle typos, slang, and ambiguous queries with remarkable accuracy.

Google's AI also powers features you might not associate with artificial intelligence. When you take a photo of text in another language and Google Translate automatically converts it, that's AI combining computer vision (to recognize text in images) with natural language processing (to translate between languages). When Google Maps predicts traffic conditions and suggests the fastest route, that's AI analyzing real-time traffic data from millions of users.

Your Digital Butler and Security Guard

Siri, Alexa, and Google Assistant represent AI's move into conversational interfaces. These systems combine several AI technologies: speech recognition (converting your voice to text), natural language understanding (figuring out what you want), knowledge retrieval (finding the right information), and speech synthesis (converting the response back to natural-sounding speech).

What makes these assistants impressive isn't that they can answer factual questions, traditional search engines could do that. It's that they can handle conversational context, ambiguous requests, and multi-step tasks. You can ask, "What's the weather like?" followed by "What about tomorrow?" and the assistant understands that the second question refers to tomorrow's weather, not some general inquiry about tomorrow.

Meanwhile, AI systems are quietly protecting you from digital threats. Your email provider uses AI to identify spam and phishing attempts, analyzing not just the content of messages but also patterns in sender behavior, email routing, and timing. Your bank uses AI to monitor your spending patterns and flag potentially fraudulent transactions. Your phone's AI can distinguish between your face and a photograph when unlocked with facial recognition.

The Social Media Experience Architect

Social media platforms use AI to curate your feed, but the sophistication goes far beyond showing you posts from friends and pages you follow. Facebook's and Instagram's algorithms consider hundreds of factors to determine what appears in your timeline: which friends you interact with most, what types of content you engage with, how much time you spend looking at different posts, and even what time of day you're most likely to be interested in various topics.

The AI learns your patterns in ways that can be almost eerily accurate. It might notice that you tend to engage more with political content on weekday mornings but prefer lighter, entertainment-focused posts in the evening. It tracks whether you're more likely to watch videos or read articles, whether you prefer content from close friends or from pages you follow, and how your engagement patterns change based on current events or seasons.

LinkedIn uses similar AI techniques to show you relevant job postings, professional updates, and networking suggestions. The system analyzes your career history, skills, connections, and engagement patterns to predict what professional content will be most valuable to you.

WHY RECOGNITION MATTERS: BUILDING CONFIDENCE IN AN AI WORLD

Understanding that you're already successfully using AI every day serves a crucial psychological purpose. It demolishes the myth that AI is too complex or advanced for regular people to understand or benefit from. You're not starting from zero—you're already an experienced AI user who simply needs to recognize and expand on skills you already have.

This recognition is empowering because it allows you to engage with and direct AI rather than just experience it. When you understand that Netflix's recommendations are based on AI analysis of your viewing patterns, you can be more intentional about rating shows and selecting content that will improve future recommendations. When you recognize that Google's search results are tailored based on AI interpretation of your query, you can refine your search terms to get better results.

From Passive Consumer to Active Collaborator

Once you recognize AI as a tool rather than magic, you can start thinking about how to collaborate with these systems more effectively. Instead of being impressed or frustrated by AI recommendations that miss the mark, you can understand that these systems learn from feedback and adjust your interactions accordingly.

For example, when Spotify's Discover Weekly playlist includes songs you don't enjoy, skipping them isn't just expressing your preference. It's training the AI to make better predictions about your musical taste. When you correct your phone's autocorrect mistakes, you're helping the system learn your personal writing style and vocabulary.

This collaborative mindset is essential as AI tools become more interactive and powerful. The people who will benefit most from AI in the coming years won't necessarily be those who understand the technical details of how these systems work. They'll be those who understand how to communicate clearly with AI, how to provide useful feedback, and how to integrate AI assistance into their workflows effectively.

Demystifying the Learning Process

One of the biggest barriers to AI adoption is the perception that you need to learn entirely new skills to benefit from these tools. The reality is that working with AI builds on communication and problem-solving skills you already have. If you can clearly explain what you want to a human assistant, you can learn to prompt AI systems effectively. If you can recognize when human-generated content is helpful or off-target, you can evaluate and improve AI-generated content.

The key insight is that AI tools are most effective when used by people who understand the domain they're working in. You don't need to understand how neural networks function to use AI writing tools effectively, but you do need to know what good writing looks like in your field. You don't need to understand computer vision algorithms to benefit from AI image generation, but you do need to know what visual style and content will serve your purposes.

THE CONFIDENCE FOUNDATION

As we move into an era where AI capabilities are expanding rapidly, the most important skill isn't technical knowledge; it's confidence. Confidence to experiment with new tools without fear of "doing it wrong." Confidence to recognize that AI is a powerful assistant, not a replacement for human judgment and creativity. Confidence to learn through hands-on experience rather than waiting until you feel you understand everything theoretically.

You've already demonstrated this confidence by successfully navigating a world filled with AI systems. Every time you've found a wonderful restaurant through Google Maps, discovered a new favorite show through Netflix, or had your phone correctly predict what you were trying to type, you've successfully collaborated with artificial intelligence.

The goal of this book isn't to turn you into an AI expert. It's to help you recognize the AI expertise you already have and expand it strategically. By the end, you'll not only understand how AI works in practical terms, but you'll also have hands-on experience using AI tools to solve real problems in your work and personal life.

The overwhelm you might have felt about AI was based on a misconception. You're not behind, struggling to catch up with a technology that has passed you by. You're already here, already using these tools successfully. Now you're ready to take the next step: using AI intentionally and strategically to enhance your productivity, creativity, and problem-solving capabilities.

The future isn't about humans versus AI. It's about humans working alongside AI to accomplish things neither could achieve alone. And that future has already begun, in your pocket, on your laptop, and in the countless small interactions that make your digital life more convenient, personalized, and effective every single day.

[2]

MEET YOUR AI TOOLKIT

CHATGPT, MIDJOURNEY & MORE

Imagine having a personal assistant who never sleeps, a creative partner who generates ideas faster than you can think them, and a research buddy who can digest information from across the web in seconds. That's not science fiction anymore, it's your Tuesday afternoon, once you know which AI tools to use and how to get started.

The AI landscape might seem overwhelming at first glance. New tools pop up weekly, each promising to revolutionize how you work, create, or think. But here's the thing: you don't need to master every AI tool that exists. You just need to understand the **core players** that can genuinely transform your daily routine, starting today.

This chapter will walk you through the essential AI tools that real people use for real work—not the flashy demos you see on social media, but the practical applications that save time, spark creativity, and solve actual problems. By the end of this chapter, you'll have your own AI toolkit set up and ready to use, all within the next 30 minutes.

THE BIG THREE: YOUR AI STARTING LINEUP

Think of AI tools like a Swiss Army knife, each tool serves multiple purposes, but some are more versatile than others. Let's start with the three tools that form the backbone of most people's AI workflows.

ChatGPT - Your Conversational Powerhouse

ChatGPT isn't just a chatbot; it's like having a knowledgeable colleague who's read everything on the internet and can help you think through problems. Created by OpenAI, this tool excels at understanding context, maintaining conversations, and adapting its responses to your specific needs.

What makes ChatGPT special isn't just that it can answer questions. It's how it can *think alongside you.* Need to brainstorm blog post ideas? It'll generate twenty unique angles you hadn't considered. Struggling with a difficult email? It can help you find the right tone, whether you need to be diplomatic, persuasive, or crystal clear.

Here's what ChatGPT does exceptionally well:

Purpose	What It Helps You Do	Example Benefit
Writing Assistance	Refine ideas, structure text, and improve clarity	Turns rough thoughts into clear emails, posts, or reports
Research Synthesis	Summarize complex topics without reading dozens of sources	Quickly get an overview of any subject
Problem-Solving	Break down complex challenges into manageable steps	Helps you think through solutions logically

Purpose	What It Helps You Do	Example Benefit
Learning Support	Explain difficult concepts in plain language	Understand topics you previously found confusing
Creative Ideation	Generate new ideas for content, projects, or brainstorming	Overcome blank-page or creative block moments

The free version gives you substantial capabilities, though you'll encounter usage limits during peak times. The paid version (ChatGPT Plus at $20/month) offers faster responses, priority access, and additional features like web browsing and image analysis.

Midjourney - Your Visual Imagination Engine

Where words fail, images speak. **Midjourney** transforms text descriptions into stunning visuals that would take human artists hours or days to create. This tool has democratized visual creation, allowing anyone to generate professional-quality images, concept art, and visual content without artistic training.

Midjourney operates through Discord, which might feel unusual at first, but this setup enables a unique community aspect where you can see other users' creations and learn from their prompts. It's like having access to a collaborative art studio where everyone's experiments inspire new possibilities.

The tool shines in several key areas:

Purpose	What It Helps You Do	Example Benefit
Marketing Materials	Create unique visuals for social media, websites, or presentations	Your content looks professional and visually distinct without hiring a designer
Concept Visualization	Turn abstract ideas into clear images	You can show others what you mean instead of *trying to explain*
Artistic Exploration	Experiment with different styles, moods, and aesthetics	Unlock new creative directions without needing technical art skills
Product Mockups	Visualize concepts before physical production	Saves time and money by testing ideas early
Storytelling	Generate illustrations that bring narratives to life	Enhances books, posts, brand stories, or presentations with engaging visuals

Midjourney operates on a credit system. The basic plan starts at $10/month and provides enough credits for most casual users. As you use more credits, you can upgrade to higher tiers that offer additional features like private galleries and commercial usage rights.

Claude - Your Thoughtful Analysis Partner

While ChatGPT excels at quick responses and broad knowledge, **Claude** (developed by Anthropic) brings a different strength: *thoughtful, nuanced analysis*. Claude tends to provide more structured, careful responses and excels at tasks requiring ethical reasoning, complex analysis, and detailed explanations.

Claude's particular strengths include:

Purpose	What It Helps You Do	Example Benefit
Document Analysis	Upload and analyze long documents (PDFs, research papers, reports)	Saves hours of reading time while still understanding key insights
Ethical Reasoning	Consider multiple perspectives and evaluate sensitive decisions thoughtfully	Produces well-balanced arguments instead of rushed or biased conclusions
Structured Thinking	Break down complicated topics into clear, logical frameworks	Makes your ideas easier to communicate, teach, or present
Code Review	Analyze programming code for structure, performance, and potential bugs	Improves code quality without needing another developer to review
Research Assistance	Synthesize information from multiple sources accurately	Helps you build well-supported conclusions faster and with greater clarity

Claude offers both free and paid tiers, with the paid version providing higher usage limits and priority access during busy periods.

THE SUPPORTING CAST: SPECIALIZED AI TOOLS WORTH KNOWING

Beyond the big three, several specialized tools excel in specific domains. You don't need all of these immediately, but knowing they exist helps you choose the right tool for specific challenges.

Notion AI - Your Smart Workspace Assistant

Notion AI integrates directly into the popular productivity platform, turning your workspace into an intelligent environment. If you already use Notion for notes, project management, or documentation, this AI enhancement feels seamless and natural.

Notion AI excels at:

Purpose	What It Helps You Do	Example Benefit
Content Generation	Create notes, drafts, outlines, and writing directly inside your workspace	Saves time switching between apps and keeps everything in one place
Database Analysis	Summarize and extract insights from tables, lists, and structured data	Quickly understand trends or priorities without manual sorting
Meeting Notes	Convert raw notes into clear action steps and follow-ups	Ensures meetings lead to decisions and clarity instead of forgotten notes
Project Planning	Suggest tasks, timelines, and structure for new or ongoing projects	Helps you organize work faster and maintain progress with less effort

Grammarly - Beyond Spell Check

Grammarly has evolved far beyond simple grammar correction. Its AI now understands context, tone, and audience, helping you communicate more effectively across different situations.

Modern Grammarly offers:

Purpose	What It Helps You Do	Example Benefit
Tone Adjustment	Match your writing style to different audiences or contexts	Makes emails, posts, and documents sound more professional or friendly when needed
Clarity Improvements	Rewrite sentences to be easier to understand	Helps your message land the first time without confusion
Plagiarism Detection	Check text against academic and online sources	Ensures originality and avoids accidental copying in professional or school work
Writing Insights	Track habits and patterns to improve your writing over time	You gradually develop clearer, stronger writing skills with less effort

Perplexity AI combines search capabilities with AI analysis, providing cited answers to complex questions. Think of it as a research assistant that not only finds information but also synthesizes it into coherent insights.

Perplexity's unique value lies in:

Purpose	What It Helps You Do	Example Benefit
Source Citation	Provide verified references for every answer	You can trust the information and easily trace where it came from
Real-Time Information	Pull data from current web content rather than static training data	Get up-to-date answers on fast-changing topics like news, markets, or tech

Purpose	What It Helps You Do	Example Benefit
Follow-Up Question Support	Ask deeper questions without losing context	Enables real "research conversations" instead of one-off answers
Visual Data Integration	Combine charts, articles, videos, and graphs into one view	Saves time by gathering multiple perspectives in one place

FREE VS. PREMIUM: MAKING SMART INVESTMENT DECISIONS

The freemium model dominates AI tools, but understanding when to upgrade requires strategic thinking about your actual usage patterns and needs.

Start Free, Upgrade Strategically

Every major AI tool offers substantial free capabilities. This isn't a limitation—it's an opportunity to explore and understand your usage patterns before investing money.

Free versions typically provide:

- Core functionality with usage limits
- Basic features sufficient for occasional use
- Opportunity to learn and develop proficiency
- Access to community resources and tutorials

Premium versions generally offer:

- Higher or unlimited usage quotas
- Faster response times and priority access
- Advanced features like plugins, integrations, or specialized models
- Commercial usage rights
- Enhanced privacy and security features

The Upgrade Decision Framework

Consider upgrading when you encounter these specific situations:

Frequency indicators: You're hitting usage limits regularly, suggesting the tool has become integral to your workflow.

Efficiency blockers: Wait times or access restrictions are impacting your productivity during important work sessions.

Feature gaps: You've identified specific premium features that would significantly improve your results or workflow.

Revenue connection: The tool directly contributes to income-generating activities, making the subscription cost easily justifiable.

Time value: Premium features save enough time to justify the cost based on your hourly value calculation.

QUICK SETUP: YOUR 30-MINUTE AI TOOLKIT INSTALLATION

Getting started with AI tools doesn't require technical expertise, but a systematic approach ensures you'll have everything configured properly and ready for productive use.

Phase One: Account Creation (10 minutes)

Step 1: ChatGPT Setup

Visit openai.com and click "Sign up." Use your primary email address —you'll receive important updates and usage notifications here. Verify your account through the email confirmation, then spend a few minutes exploring the interface. Try asking a simple question like "Explain artificial intelligence in simple terms" to get comfortable with the conversation format.

Step 2: Midjourney Preparation

Midjourney requires Discord access, so create a Discord account if you don't have one. Visit midjourney.com and click "Join the Beta." Follow the invitation link to the Midjourney Discord server. This process feels different from typical software signup, but Discord's community aspect becomes valuable as you learn from other users' creative prompts.

Step 3: Additional Tool Registration

Create accounts for Claude (claude.ai) and any specialized tools that caught your attention. Use the same email address for consistency, and consider using a password manager to keep track of your new accounts securely.

Phase Two: Basic Configuration (15 minutes)

ChatGPT Customization

Explore the settings menu to configure your preferences. Set your default language and region if different from the detected settings. Review privacy settings to understand how your conversations are handled. Create your first custom instruction, this feature allows ChatGPT to remember your preferences and context across conversations.

Midjourney Discord Navigation

Join a few of the newbie-friendly channels where you can practice without feeling overwhelmed by expert-level creations. Bookmark the official documentation and quick-start guides. Practice the basic command structure by typing "/imagine" followed by a simple description like "a peaceful mountain landscape at sunset."

Tool Integration Planning

Consider which tools might work together in your workflow. For example, you might use ChatGPT to brainstorm blog post ideas, Midjourney to create accompanying visuals, and Claude to refine the final content structure. Planning these connections early helps you develop more efficient workflows.

Phase Three: First Success (5 minutes)

Quick Win Strategy

Choose one small, immediate task where AI can provide obvious value. This might be:

- Asking ChatGPT to improve a draft email
- Generating a simple image with Midjourney for a presentation
- Using Claude to summarize a long article you've been meaning to read

Success Documentation

Keep a simple note about what worked well and what felt confusing. This becomes your personal reference guide as you explore more advanced features. Notice which tool felt most intuitive—that's probably where you should focus your initial learning efforts.

YOUR NEXT STEPS: FROM SETUP TO MASTERY

Having functional AI tools is just the beginning. The real value emerges as you develop fluency with these systems and integrate them into your regular workflow.

The Learning Curve Reality

Expect a brief adjustment period as you learn to communicate effectively with AI tools. These systems respond best to clear, specific instructions, which might feel different from how you typically request help from humans. The good news? This clarity improves your thinking and communication skills across all areas.

Start with simple, low-stakes requests and gradually increase complexity as you become more comfortable. Pay attention to which types of prompts generate the best results for your needs. Each AI tool has slightly different strengths and optimal prompt styles.

Building Your Personal AI Workflow

The most successful AI users don't just use these tools occasionally, they weave them into their regular workflows in ways that compound over time. Consider mapping out your typical work or creative processes and identifying where AI assistance could eliminate friction or enhance quality.

For instance, a small business owner might use ChatGPT for initial content ideation, Midjourney for social media visuals, and Claude for refining customer communications. A student might use Claude for research synthesis, ChatGPT for study guide creation, and Grammarly for essay polishing.

The key is starting with one clear integration point rather than trying to revolutionize everything at once. Success with AI tools comes from consistency and gradual expansion, not dramatic overnight changes.

Your AI toolkit is now ready. These tools are waiting to amplify your capabilities, accelerate your workflows, and expand your creative possibilities. The question isn't whether AI will change how you work. It's how quickly you'll harness that change to your advantage.

In our next chapter, we'll dive into the art of AI conversation, learning how to communicate with these systems to get exactly the results you need, every time.

[3]

DAY 1

YOUR FIRST CONVERSATION WITH CHATGPT

Picture this: you're standing at the edge of a swimming pool, toes curled over the concrete, knowing the water is the perfect temperature but still hesitating to take that first plunge. That's exactly how most people feel when they first encounter ChatGPT. The interface looks simple enough, just a text box with a blinking cursor, but somehow it feels intimidating.

Here's the thing: **that hesitation is completely normal**. You're about to have a conversation with artificial intelligence, something that would have been pure science fiction just a few years ago. But here's what I want you to understand from the very beginning: ChatGPT isn't going to judge your questions, roll its digital eyes at your inexperience, or make you feel foolish. It's designed to be helpful, patient, and endlessly willing to explain things in different ways until they click.

Today, we're going to have our first real conversation with ChatGPT. By the end of this chapter, you'll not only understand how to communicate effectively with AI, but you'll also have created something tangible, a short story, gotten answers to questions that have been nagging at you, or perhaps discovered a new way to approach a problem you've been wrestling with.

THE ART OF TALKING TO AI: UNDERSTANDING PROMPTS

What Is a Prompt, Really?

A **prompt** is simply the instruction or question you give to ChatGPT. Think of it as the opening line of a conversation, except this conversation partner happens to process information at lightning speed and has been trained on vast amounts of text from across the internet.

But here's where many people stumble: they treat prompts like search engine queries. They type something like "pizza recipe" and expect magic. While ChatGPT might give you a decent pizza recipe from such a basic prompt, you're missing out on its real power.

The difference between a weak prompt and a strong one is like the difference between saying "food" to a waiter versus saying, "I'd like the salmon, grilled medium, with the seasonal vegetables on the side, and could you go easy on the salt?" Both will get you fed, but one will get you exactly what you want.

The Three Pillars of Effective Prompting

Clarity is your foundation. ChatGPT is remarkably good at understanding context, but it can't read your mind. If you're thinking about writing a short story for your nephew's birthday but only type "write a story," you might get a tale about corporate espionage when what you really wanted was something about a brave little mouse.

Context is your compass. The more background information you provide, the better ChatGPT can tailor its response. Instead of "explain blockchain," try "I'm a small business owner who keeps hearing about blockchain in the news. Can you explain what it is in practical terms and whether I should care about it for my business?"

Specificity is your superpower. Rather than asking for "marketing ideas," specify: "I run a local bakery and want to attract more customers on weekday mornings. What are five creative, low-cost marketing strategies I could try this month?"

YOUR FIRST REAL CONVERSATION: A STEP-BY-STEP WALKTHROUGH

Starting Simple: The Ice Breaker

Let's begin with something straightforward. Open ChatGPT and type this prompt:

"Hi! I'm new to AI and want to understand how you work. Can you explain what you are and what you're good at, in terms a regular person would understand?"

Notice what we did here? We established context (you're new to AI), set expectations (explain in regular terms), and asked a specific question. This is infinitely better than just typing "What are you?"

When you hit enter, ChatGPT will respond with a thoughtful explanation. Read it carefully, then try a follow-up question. Maybe something like: *"That's helpful! Can you give me an example of something you're really good at and something you struggle with?"*

This is already a real conversation. You're not just getting information; you're building understanding through dialogue.

Building Confidence: The Personal Assistant Test

Now let's try something more practical. Think of a real problem or question you have. Maybe it's planning a dinner party, organizing your garage, or understanding a news topic that's been confusing you. Here's an example:

"I'm hosting a dinner party for six people next weekend. Two are vegetarian, one doesn't eat gluten, and I'm not a confident cook. Can you suggest a menu that will work for everyone and give me a realistic timeline for preparation?"

This prompt works because it includes:

- **Specific details** (six people, dietary restrictions)
- **Context** (you're not confident cooking)
- **Clear expectations** (menu suggestions and timeline)

The response you get will likely surprise you with its thoroughness and practicality. ChatGPT functions as a knowledgeable, patient assistant who doesn't judge your cooking skills or make you feel bad about waiting until the last minute to plan.

The Follow-Up Game: Going Deeper

Here's where many people miss out on ChatGPT's real value. They get their first response and move on. But the magic happens in the follow-up questions. After getting your dinner party suggestions, you might ask:

"This menu looks great, but I'm nervous about timing everything. Can you create a detailed schedule for the day of the party, starting from when I wake up?"

Or: *"The stuffed portobello mushrooms sound perfect, but I've never made them. Can you walk me through that recipe step by step, including what could go wrong and how to avoid it?"*

Each follow-up question helps ChatGPT understand your needs better and provide more targeted, useful responses.

HANDS-ON PROJECT: CREATING YOUR FIRST AI-ASSISTED SHORT STORY

The Creative Challenge

Let's create something together—a short story that showcases how AI can be a creative partner rather than a creative replacement. This exercise will help you understand prompting while producing something genuinely fun and personal.

Start with this prompt structure, but fill in your own details:

"I want to write a short story about [TOPIC/CHARACTER]. The tone should be [TONE - funny, mysterious, heartwarming, etc.]. My target audience is [AGE GROUP/RELATIONSHIP - my kids, my book club, etc.]. The story should be about [LENGTH] words. Can you help me brainstorm some plot ideas?"

For example: *"I want to write a short story about a dog who thinks he's a detective. The tone should be funny but with a little mystery. My target audience is my 8-year-old daughter who loves both dogs and mystery books. The story should be about 500 words. Can you help me brainstorm some plot ideas?"*

Collaborative Storytelling in Action

ChatGPT will give you several plot ideas. Choose one that sparks your interest, then ask for more detail:

"I love the idea about [CHOSEN PLOT]. Can you help me develop this into a more detailed outline with a beginning, middle, and end?"

Once you have an outline you're excited about, here's where it gets interesting. Don't ask ChatGPT to write the entire story. Instead, collaborate:

"Great outline! I want to write this story myself but would love your help. Can you write an engaging opening paragraph that sets up the main character and situation? Then I'll take over for the next part."

After ChatGPT provides the opening, you write the next paragraph or two. Then you might ask: *"Here's what I wrote next: [YOUR TEXT]. How does this flow with your opening? Do you have suggestions for how to transition into the main conflict?"*

The Result: Something Uniquely Yours

What you'll end up with is a story that's genuinely collaborative. ChatGPT provided structure, suggestions, and perhaps some prose, but your voice, choices, and creativity shaped the final product. This is AI as a creative partner, not a replacement.

The story you create might not win literary prizes, but it will be yours in a way that a fully AI-generated story never could be. More importantly, you'll have learned how to use AI to enhance your own capabilities rather than replace them.

THE Q&A DEEP DIVE: EXPLORING WHAT YOU'VE ALWAYS WONDERED

Turning Curiosity into Conversation

Most of us carry around questions we've always wanted to explore but never had the right person to ask. Maybe it's something you feel you "should" know by now, or a topic that interests you but seems too complex to tackle. ChatGPT excels at being that patient, knowledgeable friend who never makes you feel silly for asking.

Here's how to structure a great exploratory conversation:

"I've always been curious about [TOPIC] but found most explanations either too simple or too technical. Can you explain [SPECIFIC ASPECT] in a way that assumes I'm intelligent but not an expert? Feel free to use examples or analogies."

For instance: *"I've always been curious about how the stock market works but found most explanations either treat me like I'm five or assume I have an MBA. Can you explain how stock prices actually move up and down in a way that assumes I'm intelligent but not a finance expert? Feel free to use analogies."*

Building Knowledge Through Follow-Up Questions

The beauty of this approach is that you can immediately follow up with the specific aspects that interest or confuse you:

"That analogy about the auction house was helpful! But I'm still confused about why stock prices sometimes seem to react to news that doesn't seem directly related to the company. Can you give me some specific examples?"

Or: *"This makes sense for individual stocks, but how do things like 'the market' going up or down affect everything? What's actually happening when someone says 'the market dropped 500 points today'?"*

Each question builds on the previous answer, creating a customized learning experience that moves at your pace and focuses on what genuinely interests you.

The Power of "Explain Like I'm..." Prompts

One of the most effective prompting techniques is the "explain like I'm..." approach. You can adjust this based on your comfort level with different topics:

- *"Explain cryptocurrency like I'm someone who understands basic investing but has never touched anything digital."*
- *"Explain climate change like I'm a busy parent who wants to understand what's really happening without political spin."*
- *"Explain why people are excited about AI like I'm a small business owner wondering if this affects me."*

These prompts work because they give ChatGPT a clear framework for the complexity level and perspective you need.

KEY LESSON: AI ISN'T MAGIC, IT'S A TOOL

Demystifying the Technology

After your first real conversations with ChatGPT, you might feel a mix of amazement and slight unease. The responses can be so thoughtful, so human-like, that it's easy to attribute almost magical properties to the technology. But understanding what's actually happening will make you a much more effective user.

ChatGPT is essentially a very sophisticated pattern-matching system. It has been trained on enormous amounts of text and has learned to predict what words and phrases typically follow others in different contexts. When you ask it to write a story about a detective dog, it's drawing on patterns from thousands of stories it encountered during training, recombining elements in novel ways.

This isn't to diminish what ChatGPT can do. Pattern matching at this level produces remarkably sophisticated and useful results. But understanding this helps explain both its strengths and limitations.

Recognizing the Strengths

ChatGPT excels at:

- **Synthesizing information** from multiple sources and perspectives
- **Adapting its communication style** to match your needs and context
- **Generating creative content** by combining familiar elements in new ways
- **Breaking down complex topics** into understandable components
- **Providing patient, non-judgmental assistance** with learning and problem-solving

These strengths make it an incredibly powerful tool for learning, creativity, and productivity.

Understanding the Limitations

But ChatGPT also has clear limitations:

- **It can't browse the internet** or access real-time information
- **Its knowledge has a cutoff date** and may not include recent developments

- **It can make mistakes** and sometimes state incorrect information confidently
- **It doesn't have personal experiences** or truly understand context the way humans do
- **It can't perform actions** in the real world beyond generating text

Recognizing these limitations isn't about being negative. It's about using the tool effectively. You wouldn't expect a hammer to work as a screwdriver, and you shouldn't expect ChatGPT to do things it wasn't designed for.

The Tool Mindset: Maximum Effectiveness

The most successful ChatGPT users think of it as an incredibly versatile tool rather than a magical solution. Like any powerful tool, it requires some skill to use effectively, but that skill is learnable and doesn't require technical expertise.

Think of ChatGPT as your thinking partner. It's excellent at helping you explore ideas, organize thoughts, generate options, and work through problems. But the judgment about what to do with its suggestions? That's still entirely yours.

When you approach it this way, you'll find yourself naturally asking better questions, being more specific about what you need, and critically evaluating the responses you get. This leads to much more useful and satisfying interactions.

WHAT YOU'VE ACCOMPLISHED TODAY

Take a moment to recognize what you've actually done in this chapter. You've had meaningful conversations with artificial intelligence. You've learned to craft prompts that get you useful responses. You've either created a piece of creative writing or explored topics that interest you in depth. Most importantly, you've developed a realistic understanding of what AI can and can't do.

These aren't small achievements. Many people remain intimidated by AI or dismiss it without ever trying to understand its practical applications. You've moved past both reactions to develop actual competence.

The conversations you've had today are just the beginning. Each interaction teaches you something about how to communicate more effectively with AI, and how to integrate these capabilities into your daily life in useful ways.

Tomorrow, we'll build on these foundational skills by exploring how AI can tackle more complex, real-world problems. But for now, spend some time playing with what you've learned. Ask ChatGPT about topics that interest you. Experiment with different ways of phrasing requests. See what happens when you ask for help with actual problems you're facing.

The water's warm, and you've already taken the plunge. Now it's time to start swimming.

[4]
DAY 2
PRODUCTIVITY BOOST WITH AI

From Scattered to Streamlined: Your Second Day Journey

Remember yesterday when you first opened ChatGPT and felt that mix of excitement and uncertainty? That feeling is exactly where breakthroughs begin. Today, we're moving beyond the "wow, this is cool" phase into something far more valuable: making AI your personal productivity partner.

I've watched hundreds of people discover AI's potential for the first time. The pattern is always the same. Day one brings wonder, day two brings results, and by day three, they're wondering how they ever managed without it. You're about to experience that transformation yourself.

THE PRODUCTIVITY REVOLUTION YOU DIDN'T SEE COMING

Think about your typical Monday morning. Your inbox overflows with emails demanding responses. Three meetings need scheduling around everyone's impossible calendars. That presentation you've been dreading requires an outline, but inspiration feels miles away. Meanwhile, your to-do list grows longer while your available time shrinks.

Sound familiar? You're not alone, and more importantly, you're not stuck.

AI doesn't replace your thinking – it amplifies it. The technology excels at handling the routine cognitive tasks that drain your mental energy, freeing you to focus on work that truly requires human judgment, creativity, and relationship-building.

Sarah, a freelance marketing consultant I know, described the shift perfectly: "It's like having an incredibly capable intern who never gets tired, never takes breaks, and never needs training on basic tasks. But instead of costing me thousands per month, it costs me less than my daily coffee habit."

WRITING THAT WORKS: BEYOND GENERIC TEMPLATES

Let's start with the task that probably consumes more of your day than you'd like to admit: writing.

Email Mastery Made Simple

Most people approach AI-assisted writing wrong. They ask for "a professional email" and get back something that sounds like it was written by a robot committee. The secret lies in specificity and context.

Instead of: *"Write a professional email"*

Try: *"I need to email my biggest client about a project delay. The delay is due to unexpected feedback from their legal team, which is actually protecting them from potential issues. I want to sound accountable but also reassure them this is normal and beneficial. Keep it under 200 words and maintain a warm but professional tone."*

The difference is remarkable. The AI now has your situation, your relationship context, your desired outcome, and your communication style preferences. The result feels authentic because you've provided authentic input.

Brainstorming That Actually Produces Ideas

Traditional brainstorming often hits walls. You stare at blank pages, hoping inspiration strikes. AI transforms this process into an interactive dialogue that builds momentum rather than waiting for it.

Start with a conversation, not a command. Say: *"I'm working on a presentation about sustainable office practices for a company that's been resistant to change. Help me think through different angles to make this appealing."*

The AI might respond with environmental benefits, cost savings, employee satisfaction, competitive advantages, and regulatory compliance angles. But here's where it gets interesting. You can dig deeper: *"The cost savings angle sounds promising. What specific examples could make this concrete for skeptical executives?"*

This back-and-forth generates ideas you wouldn't have reached alone. You're not outsourcing creativity; you're creating a thinking partnership.

Meeting Notes and Summaries

Recording and summarizing meetings typically happens in one of two ways: either you frantically scribble notes while trying to participate, or you fully engage and later struggle to remember key decisions. AI offers a third option.

After your next meeting, spend five minutes doing a "brain dump" to the AI. Don't worry about organization or completeness: *"Just had a project kickoff meeting with the design team. We discussed timeline issues, budget constraints for the photography, three different layout options, concerns about the client's brand guidelines being too restrictive, and Sarah mentioned some technical limitations with the content management system."*

Then ask: *"Organize this into clear action items, decisions made, and follow-up questions that need answers."*

The AI will structure your scattered thoughts into something actionable, often catching connections and implications you missed during the conversation.

THE MINI-PROJECT: AI ORGANIZES YOUR SCHEDULE

Theory matters, but transformation happens through practice. Let's create something immediately useful: an AI-organized approach to your upcoming week.

Step 1: The Brain Dump

Open your AI tool and start a conversation like this:

"I need help organizing my schedule for next week. Let me tell you what I've got coming up, and then I'd like your help prioritizing and structuring it all."

Then list everything – meetings, deadlines, personal commitments, routine tasks, projects in various stages. Don't worry about order or importance yet. Just get it all out of your head and into the conversation.

Step 2: Context Setting

Now add the crucial context that transforms generic advice into personalized strategy:

"My biggest priority this week is [your main focus]. I work best when I have [your preferred working style – long blocks, quick tasks mixed in, mornings vs afternoons, etc.]. The things that usually derail my productivity are [your typical challenges]. I have about [realistic assessment] hours of focused work time available."

Step 3: Collaborative Organization

Ask the AI to work with you: *"Based on what I've shared, how would you structure this week? What would you prioritize, and what would you suggest moving or delegating?"*

The AI will propose a structure, but here's the important part – treat it as a starting point for refinement, not a final answer. Respond with your thoughts: *"That makes sense, but I don't think I can handle client calls on Tuesday afternoon because that's when I'm usually mentally drained from the morning meetings. Can we adjust that?"*

Step 4: Building Buffers and Contingencies

Most scheduling fails because it assumes perfect conditions. Ask your AI partner: *"What could go wrong with this schedule, and how should we build in flexibility?"*

The AI might suggest buffer time between meetings, identify potential bottlenecks, or recommend moving your most important tasks to times when you're less likely to face interruptions.

Step 5: Daily Review System

Finally, establish a simple daily check-in process: *"Create a 5-minute daily review template I can use each morning to adjust this plan based on how the previous day actually went."*

This creates an evolving, responsive system rather than a rigid schedule that crumbles at the first unexpected development.

FROM OVERWHELM TO CONTROL: THE MENTAL SHIFT

The real productivity boost from AI isn't about doing more tasks. It's about approaching tasks differently. When you know you have a capable thinking partner available, the nature of work challenges changes fundamentally.

Reducing Decision Fatigue

Every day, you make thousands of small decisions that cumulatively drain your mental energy. Should this email be formal or casual? What's the best order to tackle today's tasks? How should I structure this presentation?

AI handles these micro-decisions, preserving your mental bandwidth for choices that truly require human judgment. The result isn't just efficiency. It's sustained energy throughout your day.

Breaking Project Paralysis

Large projects often stall because they feel overwhelming in their entirety. AI excels at decomposition – taking intimidating wholes and breaking them into manageable parts.

Instead of staring at "Plan company retreat" and feeling paralyzed, you can collaborate with AI to create: venue research, budget planning, activity coordination, catering arrangements, transportation logistics, and communication planning. Each component becomes a concrete task rather than part of an overwhelming mass.

Creating Momentum Through Small Wins

AI enables what psychologists call "implementation intentions" – specific plans that connect situations to responses. Instead of vague goals like "be more organized," you can create specific, AI-supported systems that generate consistent progress.

When you successfully use AI to streamline your email responses, that success builds confidence for tackling more complex automation. When AI helps you break through writer's block once, you trust it to help again. Each small victory reinforces the productivity partnership.

REAL-WORLD APPLICATIONS: WHERE THE MAGIC HAPPENS

Let's explore how this plays out in different professional contexts, because productivity isn't one-size-fits-all.

For Small Business Owners

Running a business means juggling customer service, marketing, operations, and strategic planning simultaneously. AI becomes your Swiss Army knife for this complexity.

Use it to draft customer service responses that maintain your brand voice while addressing common concerns efficiently. Create social media content that feels authentic but doesn't consume hours of your day. Develop standard operating procedures by describing your current processes and having AI help formalize and optimize them.

Marcus, who runs a local landscaping business, uses AI to create detailed project estimates. He describes the property conditions, desired outcomes, and potential challenges to the AI, which helps him structure comprehensive proposals that address customer concerns proactively. His close rate improved by 30% because clients felt more confident in his thorough approach.

For Creative Professionals

Creativity and AI might seem contradictory, but the opposite proves true in practice. AI handles the administrative and organizational aspects of creative work, preserving mental energy for actual creation.

Freelance graphic designer Lisa uses AI to manage client communication, draft project timelines, and even brainstorm initial concept directions. "I still do all the actual design work," she explains, "but AI helps me think through problems from angles I might not have considered. It's like having a brainstorming partner who's available 24/7 and never runs out of ideas."

For Career Changers and Students

If you're learning new skills or transitioning between fields, AI becomes an invaluable study companion and skill development accelerator.

Rather than struggling through complex concepts alone, you can have AI explain difficult topics from multiple angles, create study schedules that match your learning pace, and even roleplay scenarios you might encounter in your new field.

Career changer Jennifer used AI to prepare for her transition from teaching to project management. She had AI create realistic project scenarios, help her understand industry terminology, and even practice difficult conversations she might need to have with stakeholders.

ADVANCED PRODUCTIVITY TECHNIQUES

As you become more comfortable with basic AI assistance, several advanced techniques can multiply your results.

Template Creation and Customization

Instead of starting from scratch each time, work with AI to create reusable templates that match your specific needs and communication style. Develop email templates for different client situations, meeting agenda formats for various team configurations, and project planning frameworks for your typical work types.

The key is customization. Generic templates feel impersonal, but AI-assisted templates built around your specific communication patterns and professional requirements feel authentically yours while saving significant time.

Workflow Automation Planning

AI excels at analyzing your current workflows and identifying automation opportunities. Describe your typical process for handling a recurring task, and ask AI to identify steps that could be streamlined, combined, or eliminated entirely.

You might discover that three separate tools you currently use could be replaced by one platform, or that certain approval processes could be simplified without losing necessary oversight.

Cross-Project Learning

One of AI's most powerful features is pattern recognition across different contexts. As you work with AI on various projects, it begins to understand your preferences, working style, and successful approaches. This learning compounds, making each subsequent interaction more valuable.

BUILDING YOUR AI PRODUCTIVITY SYSTEM

Success with AI productivity requires more than occasional usage – it benefits from systematic integration into your daily workflow.

Morning Planning Sessions

Start each day with a brief AI conversation about your priorities, potential challenges, and optimal task sequencing. This 5-10 minute investment often prevents hours of inefficient work later.

Midday Check-ins

Use AI to assess progress and adjust plans based on morning developments. Rather than rigidly following predetermined schedules, create responsive systems that evolve throughout the day.

Evening Reviews

End your workday by reflecting with AI on what worked well, what caused difficulties, and how tomorrow's approach might improve on today's results.

This daily rhythm creates continuous improvement in your productivity systems while building your comfort and skill with AI collaboration.

OVERCOMING COMMON OBSTACLES

Every new productivity system faces predictable challenges. Recognizing these patterns helps you navigate them successfully.

The Perfectionism Trap

Some people delay using AI for productivity because they want to learn the "perfect" way first. This approach misses the point entirely. AI productivity systems improve through use, not through theoretical study.

Start with imperfect experiments. Ask clunky questions. Accept responses that need refinement. Each interaction teaches you something about effective collaboration, building toward genuinely useful systems.

Integration Anxiety

Existing productivity habits can feel threatened by AI introduction. Rather than replacing everything at once, integrate AI gradually into your current systems.

If you already use a particular calendar method, enhance it with AI scheduling assistance. If you have established email routines, add AI drafting to specific types of messages. Build on existing foundations rather than starting over.

Consistency Challenges

Like exercise or healthy eating, productivity benefits from consistency more than perfection. Create simple, repeatable ways to engage with AI that don't require significant willpower or schedule disruption.

The goal is sustainable integration, not dramatic transformation that fades after a few weeks.

TOMORROW'S FOUNDATION

As you finish today's exploration of AI productivity, notice how your relationship with technology has already shifted. Yesterday, AI felt like an external tool. Today, it's becoming a collaborative partner.

This evolution continues tomorrow when we explore AI's creative capabilities. But the foundation we've built today – understanding AI as a thinking amplifier rather than a replacement, learning to provide context for better results, and developing systematic approaches to integration – underlies everything that follows.

Your productivity transformation isn't just about doing things faster. It's about approaching work with greater confidence, clarity, and creative capacity. When routine cognitive tasks become effortless, energy flows toward the work that truly matters: building relationships, solving complex problems, and creating value that only humans can provide.

The overwhelm you felt this morning has transformed into something else entirely – the satisfying sense that you're finally working with technology instead of despite it.

[5]

DAY 3

CREATIVITY UNLEASHED

Welcome to Day 3 of your AI journey! By now, you've probably started to see how AI can handle everyday questions and solve practical problems. Today, we're stepping into something far more exciting and, frankly, more fun: **unleashing your creative potential** with artificial intelligence as your collaborative partner.

If you've ever stared at a blank page wondering what to write, or wished you could create stunning visuals without years of design training, today's your breakthrough moment. We're going to explore how AI transforms creative work from intimidating to accessible, turning you into a content creator, artist, and innovator faster than you ever thought possible.

THE CREATIVE REVOLUTION YOU CAN JOIN TODAY

Here's something that might surprise you: AI isn't replacing human creativity—it's **amplifying** it in ways we've never seen before. Think of AI as the ultimate creative assistant, one that never gets tired, never runs out of ideas, and never judges your initial drafts.

Just last month, a small bakery owner I know used ChatGPT to write engaging social media posts that increased her Instagram followers by 300% in six weeks. A freelance graphic designer discovered he could generate initial concepts with AI art tools, then refine them with his expertise, cutting his project time in half while doubling his creative output.

These aren't tech wizards or digital natives—they're regular people who learned to see AI as a creative partner rather than a threat.

WRITING WITH AI: FROM BLANK PAGE TO PUBLISHED POST

Breaking Through Writer's Block Forever

We've all been there: cursor blinking on an empty document, deadline approaching, brain completely blank. AI transforms this frustrating experience into something almost magical. Instead of wrestling with writer's block, you're having a conversation with an incredibly knowledgeable writing partner.

Starting Strong with Strategic Prompts

The secret to exceptional AI-generated writing lies in how you frame your requests. Forget basic prompts like "write a blog post about coffee." Instead, try something like this:

"I'm writing a blog post for busy professionals who love coffee but feel guilty about their expensive coffee shop habit. I want to show them how to make café-quality drinks at home in under 5 minutes. Write an engaging introduction that acknowledges their struggle and promises a practical solution."

Notice the difference? You've given the AI context about your audience, their pain point, and your specific goal. This transforms generic content into targeted, valuable writing.

Crafting Compelling Blog Content

When working with AI on blog posts, think of yourself as the director of a creative project. The AI provides the raw material, but you shape the vision, add personal touches, and ensure the final product reflects your unique voice.

Start by outlining your post structure with the AI:

- What problem does this solve for your readers?
- What's your main argument or solution?
- What examples or stories will illustrate your points?
- How do you want readers to feel after reading this?

Then, work section by section. Ask the AI to expand on specific points, suggest compelling headlines, or provide relevant statistics and examples. The magic happens when you blend the AI's comprehensive knowledge with your personal insights and experiences.

Poetry and Creative Writing Adventures

Don't let poetry intimidate you—AI makes it surprisingly accessible. Whether you're crafting a heartfelt message for a special occasion or exploring creative expression for personal fulfillment, AI can guide you through different poetic forms and styles.

Try experimenting with specific requests: *"Write a haiku about the feeling of accomplishment after learning something new"* or *"Help me write a short poem in the style of Maya Angelou about overcoming fear."*

The beauty of AI-assisted poetry lies in its ability to offer multiple variations quickly. You can explore different emotional tones, experiment with various structures, and discover poetic techniques you might never have encountered otherwise.

Social Media That Actually Engages

Social media content often feels like shouting into the void, but AI can help you create posts that genuinely connect with your audience. The key is understanding that different platforms require different approaches, and AI excels at adapting content for specific contexts.

For **LinkedIn**, you might ask: *"Turn my recent client success story into a professional LinkedIn post that offers value to other small business owners. Keep it under 150 words and include a thought-provoking question at the end."*

For **Instagram**, try: *"Create an engaging caption for my photo of a home-cooked meal that encourages followers to share their own cooking wins. Use a warm, encouraging tone and include relevant hashtags."*

The AI understands platform conventions, optimal post lengths, and engagement strategies, giving you content that feels native to each platform while maintaining your authentic voice.

VISUAL CREATIVITY: BRINGING IDEAS TO LIFE

AI Art Tools That Democratize Design

Visual creativity was once the exclusive domain of trained artists and designers. Today, AI art generators like **Midjourney**, **DALL-E**, and **Stable Diffusion** have opened up incredible possibilities for anyone with imagination.

These tools don't require you to understand perspective, color theory, or composition—though having some basic knowledge certainly helps. Instead, they respond to detailed descriptions, allowing you to communicate your vision in words and watch it come to life visually.

Mastering the Art of Visual Prompts

Creating compelling AI art starts with learning to describe images effectively. Instead of "a beautiful landscape," try "a serene mountain lake at sunrise, with mist rising from the water, surrounded by autumn trees in vibrant oranges and reds, soft golden light filtering through the forest, photographic style."

The more specific you are about:

- **Setting and atmosphere** (time of day, weather, mood)
- **Visual style** (photography, painting, digital art, specific artistic movements)
- **Colors and lighting** (warm vs. cool tones, dramatic vs. soft lighting)
- **Composition elements** (foreground, background, focal points)

...the more likely you are to generate images that match your creative vision.

Design Applications for Real Projects

AI-generated visuals aren't just for artistic exploration—they're incredibly practical for business and personal projects. Small business owners use AI to create:

- **Social media graphics** that maintain consistent branding
- **Product mockups** for testing new concepts
- **Blog post headers** that complement written content
- **Marketing materials** for events and promotions
- **Presentation slides** that engage audiences

The key is learning to iterate. Generate multiple versions, combine elements from different images, and use AI-created visuals as starting points for further customization in simple editing tools.

Video Content and Motion Graphics Ideas

While AI video generation is still evolving, AI excels at conceptualizing video content and creating supporting materials. Use AI to:

Brainstorm video concepts: *"I want to create a 2-minute educational video explaining compound interest to college students. Suggest five different creative approaches that would keep them engaged."*

Write video scripts: AI can help structure your ideas into compelling narratives with clear beginnings, middles, and ends.

Generate storyboards: Describe each scene you envision, and AI can help you visualize the flow and pacing of your video content.

Create supporting graphics: Generate title cards, background images, and visual elements that enhance your video production.

MINI-PROJECT: CRAFTING YOUR FIRST AI-ASSISTED BLOG POST

Now for the exciting part—let's put everything together in a hands-on project that will give you immediate, practical experience with AI-powered content creation.

Step 1: Choose Your Topic and Audience

Select something you genuinely know about or care about. This could be:

- A hobby or skill you've developed
- A problem you've solved in your work or personal life
- An experience that taught you something valuable
- A topic you've been curious about and want to explore

Define your audience specifically. Instead of "everyone," think about who would genuinely benefit from reading about your topic. Are you writing for fellow parents, small business owners, people learning a new skill, or individuals facing a specific challenge?

Step 2: Create Your Content Framework

Open ChatGPT and start with a comprehensive prompt that establishes your project parameters:

"I'm writing a blog post for [specific audience] about [your topic]. The goal is to [what you want to achieve - inform, inspire, solve a problem, share an experience]. The tone should be [conversational, professional, encouraging, etc.]. Help me create an outline with 5-7 main points that would be valuable to this audience."

Review the suggested outline and modify it based on your expertise and perspective. Remember, you're the creative director—the AI is providing structure and suggestions.

Step 3: Develop Your Content Section by Section

Rather than asking the AI to write the entire post at once, work collaboratively:

For the introduction: *"Write an engaging opening paragraph that [acknowledges reader's challenge/sparks curiosity/shares a relatable moment]. Make it feel conversational and draw readers in immediately."*

For main sections: *"Expand on the point about [specific outline item]. Include practical examples and explain why this matters to my audience. Keep it around 200 words."*

For transitions: *"Suggest a smooth transition from the section about [topic A] to [topic B] that maintains reader engagement."*

Step 4: Add Your Personal Touch

This is where you transform AI-generated content into something uniquely yours:

- **Insert personal anecdotes** that illustrate your points
- **Add your own examples** from your experience or observation
- **Modify the tone** to match how you actually speak and write
- **Include specific details** that only you would know or think to mention

Step 5: Polish and Optimize

Ask the AI to help with final touches:

- *"Suggest three compelling headlines for this blog post"*
- *"What are five key takeaways readers should remember from this post?"*
- *"How can I end this post in a way that encourages readers to take action or continue learning?"*

Step 6: Review and Refine

Read your completed draft as if you were your target audience. Does it:

- Address their specific needs and interests?
- Feel authentic to your voice and perspective?
- Provide genuine value they can act on immediately?
- Flow naturally from one idea to the next?

Make adjustments based on your assessment, using the AI as a sounding board for improvements.

BUILDING CREATIVE CONFIDENCE THROUGH EXPERIMENTATION

The most important thing to understand about AI-assisted creativity is that it's iterative. Your first attempts might feel awkward or overly dependent on the AI's suggestions. That's completely normal and expected.

Start with low-stakes projects. Write social media posts, create simple graphics for personal use, or draft emails with more engaging language. As you become comfortable with the collaborative process, tackle more ambitious creative projects.

Embrace the learning curve. Each interaction with AI tools teaches you something about effective prompting, creative problem-solving, and your own preferences as a creator.

Develop your unique approach. Notice what types of prompts yield results you love, which creative directions feel most natural to you, and how you prefer to blend AI assistance with your personal input.

YOUR CREATIVE TOOLKIT EXPANDING

By the end of today, you'll have experienced something profound: the democratization of creativity through AI. Tools that once required years of training to master are now accessible through conversation and experimentation.

You're not just learning to use specific AI applications. You're developing a new creative mindset. One that sees AI as a collaborative partner, embraces experimentation over perfection, and understands that the best creative work emerges from the intersection of human insight and artificial intelligence capabilities.

Tomorrow, we'll explore how AI can enhance your professional productivity and streamline your work processes. But first, take a moment to appreciate what you've accomplished today. You've joined the ranks of creators who are shaping the future of human-AI collaboration, one creative project at a time.

The blank page will never feel quite as intimidating again.

[6]

DAY 4

DEEP LEARNING WITHOUT THE MATH

Welcome to day four of your AI journey. By now, you've dipped your toes into the world of artificial intelligence, learned about machine learning, and maybe even tried your hand at some basic AI tools. Today, we're going deeper, literally. We're talking about **deep learning**, the technology that powers the most impressive AI breakthroughs you've been hearing about.

Don't worry. I promised we'd skip the math, and I meant it. Instead of drowning you in equations and algorithms, we're going to understand deep learning the way it actually works in the real world. Think of this as your backstage pass to the most exciting show in technology.

THE BRAIN THAT ISN'T REALLY A BRAIN

Before we dive into neural networks, let me clear up the biggest misconception about deep learning. Despite what you might have heard, AI doesn't actually think like a human brain. The term "neural network" is more marketing than biology—it's inspired by how we think brains work, but the resemblance ends there.

Here's what's really happening: imagine you're teaching a child to recognize cats in photos. You'd show them hundreds of pictures, pointing out whiskers, pointy ears, and furry tails. Eventually, they'd learn to spot cats on their own. Deep learning works similarly, but instead of one child, you have millions of tiny mathematical decision-makers all working together.

What Makes It "Deep"

The "deep" in deep learning refers to layers—lots and lots of them. Think of these layers like a factory assembly line, where each station adds something to the final product. In a neural network, each layer looks for different patterns and passes its findings to the next layer.

The first layer might notice basic shapes and colors. The second layer combines these into more complex patterns like curves and corners. By the time you get to the final layers, the system can recognize entire objects, faces, or even abstract concepts. It's pattern recognition on steroids, performed at a scale that would make human pattern-seekers dizzy.

THE MAGIC BEHIND THE CURTAIN

Let's break down how neural networks actually learn, using an example everyone can relate to: spam detection.

Imagine you're sorting through your email. You quickly learn to spot spam—weird subject lines, too many capital letters, suspicious sender addresses. A neural network learns the same way, but it examines thousands of features simultaneously. It might notice that emails with the words "urgent" and "limited time" sent from addresses ending in random numbers are usually spam.

Here's where it gets interesting: the network doesn't just follow rules someone programmed. It discovers these patterns by analyzing millions of examples. Feed it enough spam and legitimate emails, and it starts recognizing subtle patterns that even humans miss. Maybe it notices that legitimate emails from banks never use certain fonts, or that real promotional emails have specific formatting patterns.

Training: The Digital Apprenticeship

Training a neural network is like running the world's most intensive apprenticeship program. You start with a network that knows absolutely nothing—it's essentially making random guesses. Then you show it example after example, constantly correcting its mistakes.

"No, that's not a cat, that's a dog."

"Yes, that email is spam."

"Close, but that's a car, not a truck."

After millions of these corrections, something remarkable happens: the network starts getting things right. Not because it understands cats or dogs or spam in any human sense, but because it's found mathematical patterns that reliably predict the right answers.

REAL-WORLD DEEP LEARNING IN ACTION

Now let's look at where deep learning is actually changing the world, and your daily life, right now.

Image Recognition: Beyond Human Ability

Remember when tagging friends in Facebook photos was a manual chore? Deep learning changed that forever. Modern image recognition doesn't just identify faces—it can tell you who's in a photo even if they're wearing sunglasses, shot from an unusual angle, or partially hidden behind someone else.

But the applications go far beyond social media. Doctors use deep learning to spot cancer in medical scans earlier than human radiologists can detect it. Self-driving cars use it to distinguish between a plastic bag blowing across the road and a child chasing a ball. Security systems can identify specific individuals in crowded airports, even when they're trying to hide their faces.

The most mind-blowing part? These systems often surpass human performance. A deep learning model trained on enough dermatology images can diagnose skin cancer more accurately than experienced dermatologists. That's not because the AI is "smarter"—it's because it's processed more examples than any human could see in multiple lifetimes.

Voice Assistants: Understanding Human Speech

Every time you ask Alexa about the weather or tell Siri to set a timer, you're using deep learning. Converting human speech into text and understanding what you mean is incredibly complex. Consider how many different ways people say "What's the weather like?"

- "How's it looking outside?"
- "Do I need an umbrella today?"
- "Is it going to rain?"
- "What should I wear?"

A traditional computer program would need explicit rules for every possible variation. Deep learning models learn to understand intent by analyzing millions of conversations. They pick up on context, regional accents, and even emotional undertones.

The breakthrough moment for voice recognition came when researchers stopped trying to program language rules and started letting neural networks discover patterns in human speech. The result? Systems that can understand you even when you're mumbling, speaking with an accent, or using slang they've never encountered before.

Translation: Breaking Down Language Barriers

Google Translate used to be a source of hilarious mistranslations. Now it's genuinely useful for real communication across language barriers. This transformation happened almost overnight when Google switched from rule-based translation to deep learning.

The old system worked like an overly literal student with a dictionary, translating word by word and following grammatical rules. Deep learning approaches translation more like a human polyglot—understanding meaning, context, and cultural nuances. It doesn't just translate words; it translates ideas.

The system learned by analyzing millions of documents that exist in multiple languages—UN proceedings, EU legislation, international news articles. By comparing how the same ideas are expressed across languages, it developed an almost intuitive understanding of how thoughts map between different linguistic systems.

YOUR FIRST AI ART ADVENTURE

Ready for some hands-on experience? Today's mini-project involves creating your first AI-generated image. This isn't just for fun—it's a perfect way to understand how deep learning transforms simple text into complex visual creations.

Choosing Your Platform

Several platforms make AI image generation accessible to beginners. DALL-E 2 from OpenAI offers high-quality results with a user-friendly interface. Midjourney creates stunning artistic images and runs through Discord. Stable Diffusion provides more control and runs locally on your computer if you're technically inclined.

For your first attempt, I recommend starting with DALL-E 2 or Midjourney. Both have intuitive interfaces and produce impressive results without requiring technical setup.

Crafting Your First Prompt

The art of AI image generation starts with learning how to communicate with the system. Unlike traditional art software where you draw what you want, AI image generators work with text descriptions called **prompts**.

Start simple: "A sunset over a mountain lake." The AI will interpret this and generate something reasonable. But here's where it gets interesting—small changes in your prompt can produce dramatically different results.

Try these variations:

- "A dramatic sunset over a crystal-clear mountain lake"
- "A peaceful sunset over a mountain lake, painted in watercolor style"
- "A sunset over a mountain lake, shot with a vintage film camera"

Notice how each addition guides the AI toward different artistic interpretations. The word "dramatic" might add more intense colors and dynamic cloud formations. "Watercolor style" changes the entire rendering technique. "Vintage film camera" adds grain and affects the color palette.

Understanding What's Happening Behind the Scenes

When you submit your prompt, the AI isn't searching through a database of existing images. Instead, it's using everything it learned from analyzing millions of pictures during training. It knows what sunsets typically look like, how mountains are usually shaped, and what makes a lake appear peaceful or dramatic.

The neural network is essentially asking itself: "Given all the sunsets, mountains, and lakes I've seen, what would an image matching this description look like?" It then generates something entirely new that fits your description.

This is why AI art can be both familiar and surprising. The AI combines elements it has learned in ways that might never have existed in its training data. Your sunset might have the color palette from a famous painting, the mountain shapes from Ansel Adams photographs, and water reflections from contemporary digital art—all blended into something completely original.

Experimenting with Style and Detail

Once you've created your first image, start experimenting with more specific prompts. AI image generators understand artistic styles, historical periods, and technical photography terms.

Try adding style descriptors:

- "...in the style of Van Gogh"
- "...as a Japanese woodblock print"
- "...like a 1950s travel poster"
- "...rendered in photorealistic detail"

Technical terms can also dramatically affect results:

- "...with shallow depth of field"
- "...dramatic lighting"
- "...wide-angle lens"
- "...golden hour lighting"

The more specific your prompt, the more control you have over the final image. But don't be afraid to leave some elements vague—sometimes the AI's interpretation of ambiguous prompts leads to the most interesting results.

THE DEEPER UNDERSTANDING

As you create more AI images, you're witnessing deep learning in action. The system generating your artwork learned from millions of images, identifying patterns in everything from color relationships to compositional techniques. It doesn't "know" what a sunset is in any human sense, but it has developed incredibly sophisticated mathematical relationships that capture the visual essence of sunsets.

Pattern Recognition at Scale

This is the core insight about deep learning: it's pattern recognition performed at a scale and complexity that humans can barely comprehend. The neural network processing your image prompt has analyzed more visual information than any human artist could study in multiple lifetimes.

When you ask for a "dramatic sunset," the AI draws upon patterns from thousands of dramatic sunsets—not by copying any specific image, but by understanding the mathematical relationships that make sunsets appear dramatic. The interplay of warm colors, the positioning of clouds, the reflection patterns on water—all of these elements are encoded as numerical relationships within the network.

Learning Without Understanding

Here's where deep learning gets philosophically interesting: these systems achieve remarkable results without understanding anything in a human sense. The AI generating your sunset image doesn't experience the emotional impact of a beautiful sunset or understand why certain color combinations are aesthetically pleasing. It has simply identified mathematical patterns that correlate with human preferences.

This distinction matters because it helps us understand both the power and limitations of deep learning. These systems can identify patterns too subtle for humans to notice and process information at impossible scales. But they lack the contextual understanding, emotional intelligence, and creative insight that humans bring to problem-solving.

BEYOND THE HYPE: REAL LIMITATIONS

While deep learning has achieved impressive results, it's important to understand its current limitations. AI image generators occasionally produce bizarre results—extra fingers on hands, impossible architecture, or objects that defy physics. These failures happen because the system is working with statistical patterns rather than true understanding.

The Data Dependency

Deep learning systems are only as good as their training data. If an AI image generator was trained primarily on photos from certain regions or cultures, it might struggle with prompts involving unfamiliar concepts. This is why understanding training data and potential biases becomes crucial as these systems become more prevalent.

The Creativity Question

The images you generate today raise fascinating questions about creativity and authorship. When you create an AI artwork, who is the real artist? You provided the creative direction and made aesthetic choices about prompts and iterations. The AI contributed technical execution and pattern synthesis. The original artists whose work contributed to the AI's training provided the visual knowledge that made generation possible.

These aren't just academic questions—they're shaping legal and ethical discussions about AI-generated content, copyright, and creative ownership.

LOOKING FORWARD: YOUR NEXT STEPS

Today's exploration of deep learning and hands-on image generation represents a significant milestone in your AI journey. You've moved beyond simply using AI tools to understanding how they work and creating original content with them.

The skills you're developing—prompt crafting, understanding AI capabilities and limitations, creative problem-solving with AI assistance—are becoming increasingly valuable across industries. Marketing professionals use AI image generation for rapid prototyping. Architects create concept visualizations. Authors generate cover art ideas.

Building Your AI Toolkit

As you continue experimenting with AI image generation, you're building practical skills in human-AI collaboration. Learning to communicate effectively with AI systems through well-crafted prompts is becoming as important as any traditional creative skill.

Consider keeping a collection of your most successful prompts and noting what made them work. Document the relationship between specific words and visual outcomes. This experiential learning will serve you well as AI tools become more sophisticated and ubiquitous.

Tomorrow, we'll explore how these same deep learning principles power language models—the technology behind ChatGPT and similar conversational AI systems. You'll discover how the pattern recognition skills you've seen in image generation apply to understanding and generating human language.

But for now, enjoy creating with your new understanding of deep learning. Every image you generate is a small miracle of pattern recognition, mathematics, and creative collaboration between human insight and artificial capability. Welcome to the future of creative technology.

[7]

DAY 5

BUILD YOUR FIRST AI PROJECT

Welcome to your graduation day. After four days of experimenting with individual AI tools, you're ready to orchestrate them together like a conductor leading a symphony. Today isn't about learning another new tool—it's about becoming an AI architect, someone who can envision a project and bring it to life by combining the digital assistants you've already met.

Think of this as your first real test drive. You've practiced parallel parking and highway merging; now it's time to plan a road trip that showcases everything you've learned. The beauty of AI projects isn't their complexity—it's how simple tools can create surprisingly sophisticated outcomes when you connect them thoughtfully.

WHY MULTI-TOOL PROJECTS CHANGE EVERYTHING

The Multiplication Effect

When you use AI tools in isolation, you get helpful results. When you combine them strategically, something magical happens: the output becomes greater than the sum of its parts. A ChatGPT conversation becomes the foundation for a Canva presentation, which becomes the script for a video, which becomes the centerpiece of a social media campaign. Each tool amplifies the others.

Consider Sarah, a small business owner who wanted to launch a new service. Instead of spending weeks creating materials from scratch, she spent one afternoon building an AI-powered launch sequence. She used ChatGPT to refine her service description and identify target customers, Canva to create professional visuals, and Grammarly to polish her website copy. The result? A cohesive marketing package that would have cost thousands to outsource.

Beyond the Beginner's Plateau

Single-tool usage keeps you in the shallow end of AI capabilities. Multi-tool projects push you into deeper waters where the real benefits live. Instead of asking "What can this tool do?" you start asking "What do I want to accomplish, and which tools can help me get there?" This shift in thinking transforms you from a user into a strategist.

The difference is profound. A user might generate a few social media posts with ChatGPT and call it done. A strategist creates a content calendar with ChatGPT, designs matching graphics with Canva, schedules posts with a social media tool, and tracks engagement with analytics—all flowing together as one cohesive system.

THREE PROJECT BLUEPRINTS TO GET YOU STARTED

Project 1: The Complete Lesson Plan (Educators and Trainers)

The Challenge: Creating engaging educational content typically requires hours of research, content development, visual design, and assessment creation. Most educators piece this together over several days, often sacrificing quality for speed.

The AI Solution: A coordinated approach that handles research, content creation, visual design, and assessment in one seamless workflow.

Step-by-Step Breakdown:

Start with **ChatGPT** as your curriculum architect. Instead of asking for a generic lesson plan, provide specific context: "I need a 45-minute lesson on financial literacy for college freshmen who have never balanced a checkbook. Most learn better through hands-on activities than lectures. Include three interactive exercises and assessment questions."

ChatGPT will generate a structured outline, but don't stop there. Ask follow-up questions to deepen the content: "What common misconceptions should I address?" and "How can I make budgeting relatable to students living on ramen budgets?" This iterative conversation creates nuanced content that generic templates never achieve.

Next, move to **Canva** for visual storytelling. Transform ChatGPT's key concepts into infographics, slide presentations, or handout materials. Search for "financial literacy" templates, but customize them with your specific examples and data points. Create a consistent visual theme that reinforces your lesson objectives—perhaps using green for "good financial habits" and red for "money traps to avoid."

For interactive elements, return to **ChatGPT** with specific requests: "Create a realistic scenario where a college student must make five financial decisions. Include tempting but poor choices alongside smart alternatives." Use these scenarios to build role-playing exercises or case study discussions.

Finally, develop assessments that match your teaching style. ChatGPT can generate quiz questions, but ask for variety: "Create ten assessment questions—three multiple choice testing basic concepts, four scenario-based short answers testing application, and three reflection questions testing critical thinking."

The Result: A comprehensive, visually appealing lesson that feels professionally designed rather than hastily assembled. Total time investment: 2-3 hours instead of 8-10 hours of traditional preparation.

Project 2: The Business Idea Accelerator (Entrepreneurs and Side-Hustlers)

The Challenge: Moving from a vague business idea to a concrete plan typically involves market research, competitive analysis, financial projections, and marketing strategy development. Traditional approaches require specialized knowledge and significant time investment.

The AI Solution: A systematic approach that validates your idea, identifies opportunities, and creates actionable next steps using AI as your business consultant team.

Phase 1: Idea Validation and Market Research

Begin with **ChatGPT** as your market research analyst. Don't just describe your business idea—provide context about your background, resources, and constraints. "I'm a working parent with two hours of evening time and $500 to invest. I want to start a meal prep service for busy families in suburban areas. I'm good at cooking but have no business experience."

Press ChatGPT for specific market insights: "What are the biggest challenges meal prep businesses face?" and "How do successful meal prep services differentiate themselves from competitors?" Follow up with targeted questions about your local market, pricing strategies, and startup requirements.

Use **Perplexity** or similar research tools to verify and expand on ChatGPT's insights. Search for recent industry reports, local competitor analysis, and regulatory requirements for food businesses in your area. This dual approach gives you both strategic thinking and factual validation.

Phase 2: Business Model Development

Return to **ChatGPT** with your refined understanding. Request specific deliverables: "Based on our discussion, create a simple business model canvas for my meal prep service. Include revenue streams, key partnerships, cost structure, and customer segments." Ask for realistic financial projections: "Estimate monthly costs and potential revenue for the first six months, assuming I start with 10 customers and grow to 30."

Don't accept the first response. Push for details: "What equipment costs am I overlooking?" and "How should I price my services compared to local competitors?" This iterative process helps identify potential problems before they become expensive mistakes.

Phase 3: Marketing and Launch Strategy

Use **ChatGPT** to develop customer personas and marketing messages. "Describe three ideal customers for my meal prep service. What problems am I solving for each? What marketing channels are most likely to reach them?" Then brainstorm specific content ideas: "Create ten social media post concepts that would appeal to busy parents interested in healthy eating."

Transfer these concepts to **Canva** for visual execution. Create a cohesive brand identity with consistent colors, fonts, and imagery. Design social media templates, business cards, and basic marketing materials that look professionally crafted rather than homemade.

For written content, use **Grammarly** to refine your website copy, email newsletters, and social media captions. The goal isn't perfect prose—it's clear, professional communication that builds trust with potential customers.

The Result: A validated business concept with concrete next steps, professional marketing materials, and realistic financial projections. You'll move from "I have an idea" to "I have a plan" in a single weekend.

Project 3: The Mini-eBook Creation System (Content Creators and Thought Leaders)

The Challenge: Publishing authoritative content traditionally requires extensive research, professional writing skills, design expertise, and publishing knowledge. Most people abandon eBook projects because the technical and creative demands seem overwhelming.

The AI Solution: A streamlined process that handles content research, writing assistance, design, and formatting through coordinated AI tool usage.

Foundation Building: Research and Outlining

Start with **ChatGPT** as your research partner and content strategist. Choose a topic you're genuinely knowledgeable about—your professional expertise, personal passion, or life experience. "I want to write a 30-page eBook about time management for freelancers who struggle with inconsistent income. I've been freelancing for three years and learned hard lessons about productivity during feast and famine periods."

Request a detailed outline: "Create a logical chapter structure for this eBook. Include specific subtopics and key points for each chapter." Push for depth: "What unique insights should I include that aren't covered in generic productivity books?" and "What common time management advice doesn't work for freelancers?"

Use your conversation to identify knowledge gaps. When ChatGPT suggests concepts you're unfamiliar with, ask for explanations and examples. This process either fills gaps in your understanding or reveals areas where additional research is needed.

Content Development: Writing with AI Assistance

Use **ChatGPT** as your writing partner, not your ghostwriter. Start each chapter by providing context: "I'm writing chapter three about managing irregular workloads. My main points are: planning for income fluctuations, batching similar tasks, and maintaining productivity during slow periods. Help me develop these ideas."

Request specific types of assistance: "Suggest three real-world examples that illustrate the challenges of irregular workloads" or "Help me explain the psychology behind feast-or-famine anxiety in simple terms." Use ChatGPT to overcome writer's block, not to replace your voice and experience.

For each chapter, write your own first draft based on your knowledge and experience. Then use **ChatGPT** to suggest improvements: "Review this chapter introduction. Is it engaging? Does it clearly preview what readers will learn?" Follow up with specific requests: "Suggest three ways to make this section more actionable" or "Help me create a better transition between these two concepts."

Professional Polish: Editing and Design

Run your completed manuscript through **Grammarly** for comprehensive editing. Don't just accept all suggestions—use Grammarly as a proofreading partner that catches errors you missed while helping you refine your writing style.

Move to **Canva** for eBook design and layout. Search for professional eBook templates that match your content tone. Customize with your own color scheme, fonts, and imagery. Create a compelling cover design, chapter dividers, and formatted interior pages that look professionally published.

Pay special attention to visual elements that support your content. Create diagrams, infographics, or charts that illustrate key concepts. Design chapter summaries, action item checklists, or resource lists that add value beyond the written content.

The Result: A professional-quality eBook that showcases your expertise while providing genuine value to readers. The entire process, from initial outline to finished product, can be completed in 10-15 focused hours spread over several days.

YOUR FIRST MINI-PROJECT: THE 90-MINUTE CHALLENGE

Now it's time for hands-on practice. Choose one project that aligns with your current needs or interests. Don't aim for perfection—aim for completion. This first project is about understanding the workflow, not creating a masterpiece.

Setting Yourself Up for Success

Choose Your Project Based on Immediate Needs: If you're a teacher, tackle the lesson plan. If you're considering a side business, explore the business accelerator. If you want to establish thought leadership, try the mini-eBook. Motivation comes from working on something you'll actually use.

Time-Box Your Work: Set a timer for 90 minutes and work steadily toward completion. This constraint forces you to focus on essential elements rather than getting lost in perfectionism. You can always refine your project later.

Prepare Your Digital Workspace: Open all necessary tools in separate browser tabs. Have a notepad (digital or physical) ready for jotting down ideas and tracking your progress. Minimize distractions by closing email, social media, and other non-essential applications.

The Step-by-Step Process

Minutes 1-15: Planning and Setup

- Define your specific project goal in one sentence
- Gather any background information or resources you'll need
- Create a simple checklist of deliverables you want to complete

Minutes 16-45: Primary Content Creation

- Use ChatGPT for initial research, outlining, or brainstorming
- Focus on getting core content completed rather than perfecting details
- Ask specific, actionable questions rather than generic requests

Minutes 46-75: Enhancement and Refinement

- Move content to design tools like Canva for visual enhancement
- Use editing tools like Grammarly for polish and professionalism
- Focus on one key improvement that makes the biggest visual impact

Minutes 76-90: Final Assembly and Next Steps

- Combine all elements into your finished deliverable
- Identify 2-3 specific improvements you could make with more time
- Save your work and note what you learned about the process

Common Pitfalls and How to Avoid Them

The Perfection Trap: Your first multi-tool project won't be perfect, and that's exactly the point. You're learning process and workflow, not creating a final product. Embrace "good enough" as your standard for this practice round.

Tool Switching Overload: Resist the urge to try every available tool. Stick to 2-3 AI tools maximum for your first project. Depth beats breadth when you're learning coordination skills.

Scope Creep: As you work, you'll think of additional features and improvements. Write these ideas down for later rather than expanding your current project. Completion builds confidence; endless expansion builds frustration.

MEASURING YOUR SUCCESS

Immediate Indicators

Completion Over Perfection: Did you finish something tangible within your time limit? A completed first draft beats an unfinished masterpiece every time.

Process Confidence: Do you feel more confident about coordinating multiple AI tools? Can you envision tackling similar projects in the future?

Problem-Solving Skills: When you encountered obstacles, did you find solutions using AI tools rather than abandoning the project?

Long-Term Impact Metrics

Application Frequency: How often do you use multi-tool workflows in your daily work or personal projects? Regular use indicates genuine integration.

Project Complexity: Are you tackling increasingly sophisticated projects? Growth comes from gradually expanding what you're willing to attempt.

Time Efficiency: Are you completing similar projects faster than before? Speed improvements suggest mastery of the coordination process.

LOOKING BEYOND YOUR FIRST PROJECT

Building a Personal AI Toolkit

Every successful project teaches you something about tool selection and workflow optimization. Start keeping notes about what works well for different types of projects. Which tool combinations create the smoothest workflows? Where do you consistently encounter friction?

Over time, you'll develop personal preferences and shortcuts that make you more efficient. Some people discover they prefer starting with visual design and working backward to content. Others find that research-heavy projects flow better with extensive ChatGPT conversations followed by rapid execution in other tools.

Expanding Your Capabilities

Once you're comfortable with basic multi-tool projects, start exploring more sophisticated integrations. Look into tools that offer automation features, like Zapier, which can connect your AI tools into semi-automated workflows. Consider learning about AI tools that specialize in your specific industry or interests.

The goal isn't to use every available AI tool—it's to build increasingly powerful solutions to problems you actually face. Each project should make your next project easier and more ambitious.

Sharing and Learning

Consider sharing your completed projects with friends, colleagues, or online communities. Not only does this provide valuable feedback, but it also helps you articulate what you've learned. Teaching others solidifies your own understanding while building your reputation as someone who can get things done with AI.

Join online communities focused on AI tool usage and project sharing. Seeing how others approach similar challenges will give you ideas for improving your own workflows.

YOUR AI PROJECT JOURNEY STARTS NOW

Today's challenge isn't about mastering every nuance of multi-tool coordination. It's about proving to yourself that you can envision a project and bring it to life using AI tools as your creative partners. The specific project you choose matters far less than the confidence you'll gain from completing it.

Remember, every expert was once a beginner who decided to start. Your first AI project won't be perfect, but it will be yours—created through your vision, enhanced by AI capabilities, and completed through your persistence.

Set that timer, choose your project, and dive in. In 90 minutes, you'll have tangible proof that you're no longer just experimenting with AI tools—you're building with them.

[8]

DAY 6

SMARTER, SAFER AI USE

Welcome to day six of your AI journey. By now, you've probably experimented with several AI tools and felt that mixture of excitement and wonder that comes with watching artificial intelligence work its magic. Maybe you've generated stunning artwork, crafted marketing copy in minutes, or had fascinating conversations with chatbots that seemed almost human.

But here's the thing about magic tricks—once you learn how they work, you realize there's always more happening behind the curtain than meets the eye. AI is no different. Today, we're pulling back that curtain to reveal both the incredible capabilities and the very real limitations of artificial intelligence. Think of this chapter as your reality check and safety guide rolled into one.

You wouldn't drive a car without understanding its blind spots, and you shouldn't use AI without understanding where it might lead you astray. The goal isn't to scare you away from these powerful tools, but to help you use them more effectively and responsibly. By the end of today, you'll have the knowledge to spot AI's quirks, verify its outputs, and navigate the ethical landscape with confidence.

UNDERSTANDING AI'S FUNDAMENTAL LIMITATIONS

The Hallucination Problem: When AI Makes Things Up

Let's start with one of AI's most notorious quirks: **hallucination**. No, your chatbot hasn't started seeing pink elephants. In AI terms, hallucination refers to when artificial intelligence confidently presents information that's simply wrong, made-up, or nonsensical, all while sounding completely authoritative about it.

Picture this scenario: You ask ChatGPT about the annual rainfall in Timbuktu, and it responds with precise statistics, complete with citations to scientific journals. The numbers look legitimate, the sources sound credible, and the confidence level is unwavering. There's just one problem, half of those journals don't exist, and the statistics were essentially fabricated on the spot.

This happens because AI language models don't actually "know" facts in the way humans do. Instead, they're pattern-matching machines that predict what text should come next based on their training data. When they encounter a question about something they haven't seen enough examples of, they might fill in gaps by combining partially remembered information in ways that create plausible-sounding but incorrect answers.

The hallucination problem becomes particularly tricky in several scenarios:

Academic and Research Contexts: AI might cite non-existent studies or attribute quotes to the wrong people. A marketing professional recently discovered this when an AI tool claimed a famous business quote came from Steve Jobs, when it actually originated from a motivational poster company in the 1990s.

Technical Specifications: When asked about specific product features or technical details, AI might confidently describe capabilities that don't exist. A small business owner told me how an AI assistant convinced her that a particular software had integration features it actually lacked, leading to a costly purchasing mistake.

Current Events: Since most AI models have knowledge cutoffs, they might make up recent news or provide outdated information while presenting it as current. This creates obvious problems for anyone relying on AI for up-to-date information.

The solution isn't to avoid AI entirely, but to develop what I call "healthy skepticism." Treat AI outputs as educated first drafts rather than final truths, especially when the information is specific, recent, or consequential.

The Bias Mirror: How AI Reflects Human Prejudices

AI systems learn from human-created content, which means they inherit all our collective biases, stereotypes, and blind spots. It's like training someone by having them read everything ever written on the internet—they're going to pick up both the wisdom and the prejudices contained within that vast collection of human thought.

These biases manifest in subtle but significant ways across different applications:

Image Generation: AI art tools might consistently depict doctors as men and nurses as women, or default to showing executives as white when no race is specified. These patterns reflect historical biases in their training data but can perpetuate stereotypes when left unchecked.

Language Processing: Resume-screening AI has been found to favor candidates with traditionally male names or penalize applicants from certain neighborhoods. Writing assistants might suggest different tones or vocabulary based on perceived gender or cultural background.

Content Recommendations: AI systems might reinforce echo chambers by showing people content that confirms their existing beliefs, or make assumptions about interests based on demographic information.

Understanding bias in AI isn't about pointing fingers or assigning blame —it's about recognizing that these tools are mirrors reflecting human society, complete with all our imperfections. The key is approaching AI outputs with awareness of these potential biases and actively seeking diverse perspectives when making important decisions.

Context Gaps: When AI Misses the Big Picture

AI excels at processing vast amounts of information quickly, but it often struggles with context—the subtle background knowledge that humans take for granted. Think of AI as having incredible memory but limited wisdom about how that information fits into the real world.

Consider a simple example: If you ask an AI to write a professional email declining a meeting, it might produce a perfectly polite and grammatically correct response. But it won't know whether this is a meeting with your biggest client (requiring extra diplomacy) or a routine internal check-in (where brevity might be preferred). It can't see the relationships, politics, or cultural nuances that would influence how a human would craft that same message.

This context limitation shows up frequently in creative work. An AI might generate a marketing campaign that's technically brilliant but culturally tone-deaf, or suggest business strategies that ignore crucial industry dynamics. It's particularly noticeable in fields requiring emotional intelligence, cultural sensitivity, or deep understanding of human motivations.

The solution is to provide AI with rich context upfront and to review its outputs through the lens of your specific situation, relationships, and goals.

FACT-CHECKING AND VERIFICATION STRATEGIES

The Three-Source Rule

When AI provides factual information, especially statistics, historical claims, or technical details, implement what researchers call the **three-source rule**. Don't accept any significant fact unless you can verify it through three independent, credible sources.

Here's how to apply this systematically:

Primary Verification: Start with authoritative sources in the relevant field. For medical information, check with established medical institutions or peer-reviewed journals. For business data, consult industry reports or government statistics. For historical facts, reference academic institutions or established historical societies.

Cross-Reference Check: Look for the same information from sources that operate independently of each other. If AI claims a particular business strategy increased profits by 40%, don't just find one case study—look for multiple examples or broader research validating that approach.

Red Flag Recognition: Be especially cautious when AI provides information that seems too convenient, confirms your existing beliefs perfectly, or makes claims that would be significant if true but you haven't heard before.

Smart Search Strategies for Verification

Modern search engines can be your best allies in fact-checking AI outputs, but you need to search strategically:

Quote Search Technique: If AI attributes a quote to someone, search for the exact quote in quotation marks plus the person's name. Real quotes typically appear across multiple reputable sources with consistent attribution.

Reverse Detail Search: Take specific details from AI outputs—dates, names, statistics—and search for them independently. Fabricated details often return no results or only appear in other AI-generated content.

Source Chain Verification: When AI cites sources, don't just check if the source exists—verify that the source actually contains the information being attributed to it. Many AI hallucinations involve real sources being cited for information they don't actually contain.

Timeline Reality Checks: For current events or recent developments, cross-reference the timeline. If AI discusses a 2024 event but you're reading this in 2023, that's an obvious red flag about the information's accuracy.

Building Your Personal Verification Toolkit

Create a collection of trusted sources for different types of information verification:

News and Current Events: Establish relationships with reputable news sources across the political spectrum. Reuters, Associated Press, BBC, and NPR often provide relatively balanced coverage of breaking news.

Academic and Research Information: Google Scholar, university databases, and established research institutions provide reliable access to peer-reviewed information. Many universities offer public access to their research databases.

Technical and Professional Information: Industry associations, professional organizations, and government agencies often maintain authoritative information in their respective fields.

Business and Economic Data: Government statistics agencies, established financial news sources, and reputable business research firms provide reliable economic and business information.

The key is having these resources bookmarked and ready to use, rather than scrambling to find credible sources when you need to verify something important.

AVOIDING DANGEROUS OVERRELIANCE

Recognizing the Automation Trap

There's a seductive quality to AI that can lead to what psychologists call "automation bias"—the tendency to over-rely on automated systems and stop thinking critically about their outputs. It's the same phenomenon that causes people to drive into lakes because their GPS told them to, but applied to the realm of information and decision-making.

This overreliance becomes particularly dangerous in several scenarios:

High-Stakes Decision Making: Using AI to make important business decisions, medical choices, or legal judgments without proper human oversight can have serious consequences. AI should inform these decisions, not make them.

Creative Authenticity: While AI can spark creativity and help overcome writer's block, becoming too dependent on it for creative work can lead to generic, homogenized output that lacks personal voice and authentic perspective.

Learning and Skill Development: Students who use AI to complete assignments without engaging with the underlying concepts miss opportunities to develop critical thinking skills. Similarly, professionals who rely too heavily on AI for tasks within their expertise may find their skills atrophying over time.

Maintaining Human Judgment

The goal isn't to second-guess every AI output, but to maintain what I call "intelligent oversight"—staying actively engaged with AI outputs rather than passively accepting them.

The Reasonableness Test: Before acting on AI recommendations, ask yourself: "Does this make sense given what I know about this situation?" Trust your instincts when something feels off, even if the AI sounds confident.

The Consequence Assessment: Consider what would happen if the AI information is wrong. For low-stakes situations, you might accept more uncertainty. For important decisions, invest more time in verification.

The Expertise Filter: In areas where you have knowledge and experience, use that background to evaluate AI outputs. In areas where you lack expertise, seek input from qualified humans before making significant decisions based on AI recommendations.

Creating Healthy AI Workflows

Develop systematic approaches that leverage AI's strengths while maintaining appropriate human oversight:

Draft and Refine Process: Use AI to create initial drafts or generate ideas, then apply your judgment, expertise, and contextual knowledge to refine and improve the outputs.

Collaborative Verification: When working on important projects, have colleagues or experts review AI-generated content, especially in areas where accuracy is crucial.

Iterative Improvement: Rather than accepting the first AI output, engage in back-and-forth dialogue to refine and improve results. Ask follow-up questions, request clarifications, and provide additional context.

Documentation Habits: Keep track of which AI tools you used for what purposes, especially in professional contexts. This creates accountability and helps you identify patterns in AI performance over time.

ETHICAL GUIDELINES FOR DIFFERENT USER GROUPS

For Students: Academic Integrity in the AI Age

The integration of AI into education creates new opportunities for learning enhancement alongside new challenges for academic integrity. The key is developing approaches that support genuine learning while maintaining scholarly honesty.

Disclosure and Transparency: Many educational institutions are developing AI use policies that require students to disclose when and how they used AI assistance. Even without formal requirements, transparency about AI use demonstrates integrity and helps educators understand how these tools are being integrated into the learning process.

Learning-Focused Applications: Use AI to support understanding rather than replace thinking. This might involve asking AI to explain difficult concepts in different ways, generate practice problems, or help outline complex topics. The goal should be enhancing comprehension, not bypassing the mental work of learning.

Original Thinking Development: While AI can help generate ideas and provide feedback, ensure that your core arguments, analysis, and conclusions reflect your own thinking. AI should be a thinking partner, not a thinking replacement.

Subject-Appropriate Use: Different academic disciplines have varying relationships with AI tools. A creative writing class might encourage AI experimentation for inspiration, while a philosophy course might focus on developing original argumentation skills. Understanding these contextual differences helps guide appropriate use.

Consider developing a personal AI ethics statement for your studies—a brief document outlining how you'll use these tools to support rather than replace your learning goals.

For Professionals: Workplace Ethics and Transparency

Professional use of AI involves balancing efficiency gains with ethical responsibilities to employers, clients, and colleagues.

Client and Employer Transparency: Many professional contexts require disclosure when AI tools contribute to client deliverables. This is particularly important in fields like law, medicine, consulting, and financial services, where clients expect human expertise and judgment.

Quality Assurance Responsibilities: Professional standards typically require experts to stand behind their work. This means thoroughly reviewing and taking responsibility for any AI-generated content that goes out under your name or professional credentials.

Competitive and Confidentiality Considerations: Be mindful of what information you input into AI systems, particularly regarding proprietary business information, client confidential data, or trade secrets. Many AI platforms use inputs to improve their models, potentially exposing sensitive information.

Skill Development Balance: While AI can increase productivity, maintain and develop your core professional competencies. The goal should be augmenting your capabilities rather than replacing the expertise that makes you valuable in your field.

Team and Workplace Integration: Consider how your AI use affects colleagues and team dynamics. Open communication about AI tools and their applications can help teams develop consistent approaches and avoid confusion or redundancy.

For Creatives: Authenticity and Artistic Integrity

Creative professionals face unique challenges in integrating AI tools while maintaining artistic authenticity and professional integrity.

Voice and Style Preservation: While AI can help overcome creative blocks and generate ideas, developing and maintaining your unique creative voice remains essential. Use AI as a collaborator or inspiration source rather than a replacement for your creative perspective.

Attribution and Credit: Be transparent about AI's role in your creative process, especially when sharing or selling creative work. This transparency helps audiences understand your creative process and maintains trust in your artistic brand.

Client Expectations Management: If you're working with clients, establish clear expectations about AI use upfront. Some clients may specifically want human-only creative work, while others may be interested in AI-enhanced approaches. Clear communication prevents misunderstandings.

Artistic Growth and Development: Ensure that AI use supports rather than stunts your creative development. The goal should be expanding your capabilities and exploring new creative possibilities, not becoming dependent on AI for basic creative tasks.

Copyright and Ownership Considerations: Stay informed about evolving legal frameworks around AI-generated content ownership and copyright. These areas of law are developing rapidly, and creative professionals need to understand the implications for their work.

Universal Ethical Principles

Regardless of your role, several ethical principles should guide AI use:

Harm Prevention: Avoid using AI in ways that could harm individuals or groups, whether through spreading misinformation, perpetuating biases, or violating privacy.

Truthfulness: Be honest about AI's role in your work and its limitations. Don't present AI-generated content as having capabilities or certainties it doesn't possess.

Respect for Others: Consider how your AI use affects other people—colleagues, clients, audiences, or society more broadly. Aim for applications that benefit rather than exploit or deceive others.

Continuous Learning: Stay informed about AI developments, limitations, and best practices. The field evolves rapidly, and responsible use requires ongoing education.

Personal Responsibility: Remember that you remain responsible for the outputs and consequences of AI tools you use, regardless of how sophisticated or autonomous they appear.

BUILDING SUSTAINABLE AI HABITS

As we wrap up today's deep dive into responsible AI use, it's worth acknowledging that developing these habits takes time and practice. You don't need to become an AI expert overnight, but you do need to remain thoughtfully engaged with these powerful tools.

The most successful AI users I know have developed what I call "intelligent caution"—they're enthusiastic about AI's possibilities while remaining appropriately skeptical of its outputs. They verify important information, maintain their core skills and expertise, and stay transparent about their AI use.

Start small with these practices. Pick one verification technique to implement this week. Choose one area where you'll commit to maintaining human oversight. Establish one transparency habit in your professional or creative work. Building sustainable practices gradually is more effective than trying to revolutionize your approach all at once.

Remember, the goal isn't to limit AI's potential in your life—it's to unlock that potential responsibly and effectively. By understanding AI's limitations, verifying its outputs, and using it ethically, you're not just protecting yourself from potential pitfalls. You're also setting yourself up to get the most value from these remarkable tools while contributing to their responsible development and use across society.

Tomorrow, we'll explore how to put all these pieces together as you develop your personal AI toolkit and long-term strategy. But for today, take pride in developing the critical thinking skills that will serve you well throughout your AI journey.

The future belongs to people who can work effectively with AI while maintaining their human judgment, creativity, and ethical standards. You're well on your way to becoming one of them.

[9]

DAY 7

GRADUATION DAY - YOUR FINAL CHALLENGE

"The expert in anything was once a beginner."

- HELEN HAYES

WELCOME TO YOUR AI GRADUATION

Take a moment to congratulate yourself. Seven days ago, you might have looked at AI with a mixture of curiosity and confusion, wondering if this whole artificial intelligence thing was just another tech trend that would pass you by. Today, you're ready to tackle a comprehensive project that combines everything you've learned.

This isn't just about reaching the finish line—it's about recognizing how dramatically your relationship with technology has shifted in just one week. You've moved from being an observer of the AI revolution to becoming an active participant, someone who can harness these powerful tools to solve real problems and create meaningful results.

YOUR JOURNEY: FROM ZERO TO AI-CONFIDENT

Where You Started

Remember Day 1? You might have been someone who:

- Felt overwhelmed every time AI came up in conversation
- Wondered if you were "too old" or "not technical enough" to understand this stuff
- Assumed you'd need months of study or expensive courses to make any progress
- Worried that AI was either going to steal jobs or was just overhyped nonsense

Where You Stand Today

Now, after six days of hands-on learning, you possess a toolkit of AI skills that would have seemed magical just a week ago:

Writing and Communication: You can generate compelling content, refine your ideas, overcome writer's block, and adapt your tone for different audiences. Whether it's crafting professional emails, brainstorming marketing copy, or developing presentation materials, you understand how to collaborate with AI to enhance your written communication.

Creative Problem-Solving: You've discovered how AI can serve as your creative partner, helping you explore new ideas, generate alternatives when you're stuck, and approach challenges from fresh angles. You know how to prompt AI for inspiration while maintaining your unique perspective and voice.

Research and Learning: Complex topics no longer feel insurmountable. You can use AI to break down difficult concepts, create personalized learning plans, and find the specific information you need without drowning in search results.

Visual Creation: Even if you've never considered yourself artistic, you now understand how to work with AI image generators to create compelling visuals for presentations, social media, or personal projects.

Productivity Enhancement: You've learned to automate routine tasks, organize information more effectively, and streamline workflows that used to eat up hours of your time.

Strategic Planning: Whether for business or personal goals, you can now use AI to help analyze situations, generate options, and think through complex decisions more thoroughly.

This transformation didn't happen by accident. Each day built upon the last, creating a foundation of practical skills and, perhaps more importantly, confidence in your ability to learn and adapt to new technologies.

YOUR FINAL CHALLENGE: CHOOSE YOUR PATH

Today's challenge is different from the focused exercises you've completed over the past six days. Instead of following a specific script, you're going to design and execute a complete project that demonstrates your new AI capabilities. This project should be something meaningful to you—not just a demonstration of technical skill, but a solution to a real problem or opportunity in your life.

The Three Pathways

Choose the pathway that best aligns with your current needs and interests. Each option is designed to showcase different aspects of your AI toolkit while delivering genuine value.

PATHWAY 1: THE PRODUCTIVITY POWERHOUSE

Ideal for: Small business owners, freelancers, busy professionals, or anyone who feels overwhelmed by daily tasks and wants to reclaim their time.

Your Mission

Create a comprehensive productivity system that combines multiple AI tools to solve a specific workflow challenge you face regularly. This isn't about replacing human judgment—it's about eliminating friction and creating more space for the work that truly matters.

Project Components

Phase 1: Problem Identification and Analysis

Start by clearly defining your productivity challenge. Use AI as your thinking partner to analyze the situation thoroughly:

"I spend too much time on [specific task]. Help me break down exactly why this takes so long and identify the steps that could be streamlined or automated."

Common challenges might include:

- Email management and response
- Social media content planning
- Customer inquiry handling
- Meeting preparation and follow-up
- Project status reporting
- Research and information gathering

Phase 2: Solution Design

Work with AI to design a multi-step solution. This might involve:

Creating templates for common communications, developing standard operating procedures for routine tasks, setting up automated information gathering processes, or designing decision trees for handling frequent scenarios.

Phase 3: Implementation and Testing

Build out your complete system using the AI tools you've learned. Document each component so you can replicate and refine the process over time.

Phase 4: Results Measurement

Track your time savings and quality improvements over the next week. Use AI to help you analyze the data and identify further optimization opportunities.

Sample Project: Streamlined Client Communication

Let's say you're a freelance consultant who spends hours each week responding to similar client questions and managing project updates. Your AI-powered solution might include:

A comprehensive FAQ database with personalized responses for each client type, email templates that adapt based on project status and client communication preferences, an automated project status update system that generates weekly reports, and a client onboarding sequence that anticipates common questions before they're asked.

PATHWAY 2: THE CREATIVE CATALYST

Perfect for: Writers, artists, marketers, content creators, or anyone looking to expand their creative capabilities and explore new forms of expression.

Your Mission

Develop a complete creative project that showcases how AI can enhance rather than replace human creativity. The goal is to create something you're genuinely proud of—something that reflects your vision while demonstrating the power of human-AI collaboration.

Project Components

Phase 1: Creative Vision Development

Use AI to help clarify and expand your creative vision. Don't just ask for ideas—engage in a deep creative conversation:

"I want to create something that explores [your theme/interest]. Help me understand different ways to approach this topic and identify what makes my perspective unique."

Phase 2: Multi-Modal Creation

Combine different AI capabilities to create a rich, multi-layered project. This might involve:

Written content (articles, stories, scripts, poetry), visual elements (generated images, style references, mood boards), strategic components (marketing plans, audience analysis, distribution strategies), or interactive elements (presentations, social media campaigns, educational materials).

Phase 3: Refinement and Polish

Use AI to help refine your work, but maintain creative control. Focus on enhancing clarity, impact, and professional presentation while preserving your authentic voice.

Phase 4: Launch and Reflection

Share your creation with others and use AI to help you analyze feedback and plan future creative projects.

Sample Project: Personal Brand Story Campaign

Imagine you're developing a personal brand around your expertise in sustainable living. Your AI-enhanced creative project might include:

A compelling origin story that connects your personal journey to your professional mission, a series of educational infographics that make complex environmental topics accessible, social media content that balances information with inspiration, a lead magnet (like an e-book or course outline) that demonstrates your expertise, and a content calendar that ensures consistent, engaging communication with your audience.

PATHWAY 3: THE LEARNING ACCELERATOR

Best for: Career changers, students, professionals entering new fields, or lifelong learners who want to master complex subjects efficiently.

Your Mission

Design and complete an intensive learning project on a topic that's important to your personal or professional goals. Use AI to create a personalized curriculum, generate learning materials, and develop practical applications of your new knowledge.

Project Components

Phase 1: Learning Objective Design

Work with AI to define clear, measurable learning goals and create a realistic timeline:

"I need to understand [specific topic] well enough to [specific application] within [timeframe]. Help me break this down into manageable learning objectives and identify the most efficient learning sequence."

Phase 2: Curriculum Creation

Develop a comprehensive learning plan that includes:

Foundational concepts and advanced applications, practical exercises and real-world case studies, self-assessment tools and progress checkpoints, and resources for continued learning beyond this project.

Phase 3: Active Learning Implementation

Use AI to support different learning modalities. Create summaries and explanations, generate practice questions and scenarios, develop memory aids and mnemonics, and simulate conversations with experts in the field.

Phase 4: Knowledge Application

Demonstrate your learning by creating something tangible: a presentation, analysis, proposal, or practical solution that showcases your new understanding.

Sample Project: Digital Marketing Mastery

Suppose you're a traditional business owner who needs to understand digital marketing. Your AI-supported learning project might include:

A comprehensive overview of digital marketing channels with specific relevance to your industry, analysis of successful campaigns from businesses similar to yours, a detailed marketing plan for your own business with budget considerations and timeline, templates and checklists for ongoing campaign management, and a presentation to your team or stakeholders demonstrating your new strategic capabilities.

EXECUTING YOUR CHOSEN PROJECT

Time Management and Realistic Expectations

You have one day to complete your project, but that doesn't mean rushing through or accepting mediocre results. Instead, focus on creating something substantial but achievable. Use the **80/20 principle**—identify the 20% of components that will deliver 80% of the value, and execute those excellently.

Quality Over Complexity

A simple project executed with attention to detail and genuine thought will be far more valuable than a complex project that feels rushed or superficial. Your goal is to demonstrate mastery of AI tools while creating something genuinely useful.

Documentation and Reflection

As you work through your project, document your process. Note which AI tools worked best for different tasks, what prompting strategies were most effective, and where you encountered challenges. This documentation will be invaluable as you continue developing your AI skills beyond this week.

THE DEEPER TRANSFORMATION

Beyond Tool Mastery

While you've certainly learned to use AI tools effectively, something more profound has happened during this week. You've developed **technological confidence**—the belief that you can learn and adapt to new digital tools rather than being overwhelmed by them.

This confidence extends far beyond AI. You've proven to yourself that you can:

- Break down complex topics into manageable pieces
- Experiment with new technologies without fear of failure
- Integrate digital tools into existing workflows
- Maintain your authentic voice while leveraging technological assistance
- Approach problems strategically rather than reactively

The Mindset Shift

Perhaps most importantly, you've shifted from a **fixed mindset** about technology to a **growth mindset**. Instead of "I'm not a tech person," you now think "I can learn whatever technology I need." This single change in perspective will serve you well as the technological landscape continues to evolve.

Continuous Learning Framework

You've also developed a framework for continuous learning that you can apply to any new technology or skill:

1. **Start with clear, specific goals** rather than vague intentions to "learn about" something
2. **Focus on practical application** from day one rather than getting lost in theory
3. **Build confidence through small wins** before attempting complex projects
4. **Document your learning process** so you can replicate and improve your approach
5. **Connect new skills to existing knowledge** to accelerate understanding and retention

YOUR ONGOING AI JOURNEY

Immediate Next Steps

Completing today's project is just the beginning. Here's how to maintain momentum:

Week 2-4: Depth and Specialization

Choose one or two AI applications that showed the most promise during your week of learning. Spend additional time mastering these specific use cases. Join online communities focused on these applications, follow expert practitioners, and experiment with advanced techniques.

Month 2-3: Integration and Optimization

Focus on seamlessly integrating AI tools into your regular workflows. The goal is to make AI assistance feel natural rather than like an additional task. Create personal documentation of your most effective prompts and processes.

Month 4-6: Innovation and Leadership

Begin sharing your AI knowledge with others. This might involve training colleagues, writing about your experiences, or helping friends and family members get started with AI tools. Teaching others will deepen your own understanding and keep you motivated to continue learning.

Staying Current

The AI landscape evolves rapidly. Develop sustainable habits for staying informed:

- Follow a few high-quality AI news sources or newsletters
- Set aside time each month to experiment with new tools or features
- Connect with others who are integrating AI into similar work or interests
- Attend local meetups or online workshops when possible

Advanced Applications

As your comfort level increases, consider exploring more sophisticated applications:

- **Workflow Automation**: Connect AI tools with other software you use regularly
- **Custom Solutions**: Learn to create more specialized AI applications for your specific needs
- **Collaborative AI**: Explore how AI can enhance team projects and group decision-making
- **Strategic Planning**: Use AI for longer-term strategic thinking and scenario planning

THE BIGGER PICTURE: YOUR ROLE IN THE AI FUTURE

Contributing to Positive AI Development

As you become more skilled with AI tools, you also become part of shaping how these technologies develop and integrate into society. Your thoughtful, ethical use of AI contributes to positive examples of human-AI collaboration.

Consider how you can:

- **Model responsible AI use** by maintaining transparency about when and how you use AI assistance
- **Support inclusive AI development** by sharing feedback about tools that work well for non-technical users
- **Promote human-centered AI applications** that enhance rather than replace human capabilities
- **Advocate for AI literacy** by helping others overcome their initial barriers to adoption

The Long-Term View

The skills you've developed this week position you well for a future where AI becomes increasingly integrated into professional and personal life. Rather than feeling threatened by or disconnected from these changes, you now have the foundation to adapt and thrive.

Remember that AI technology will continue advancing rapidly, but the fundamental principles you've learned—clear communication with AI systems, strategic thinking about tool selection, and maintaining human judgment and creativity—will remain relevant regardless of how the specific tools evolve.

COMPLETING YOUR FINAL PROJECT

Final Quality Check

Before considering your project complete, use AI to help you evaluate your work:

"Review this project and help me identify areas where I could improve clarity, impact, or professional presentation. Also help me recognize what I've done particularly well."

Sharing and Celebrating

Consider sharing your completed project with someone who supported your AI learning journey this week. Whether it's a family member, colleague, or friend, showing what you've accomplished serves multiple purposes: it reinforces your own learning, demonstrates the practical value of AI tools, and might inspire others to begin their own AI learning journey.

Looking Forward

As you complete your final project, take a moment to envision how you'll use these AI skills over the coming months. Set specific goals for continued development and integration of AI tools into your regular routines.

Your week of AI learning represents more than just acquiring new technical skills—it represents your successful adaptation to a rapidly changing technological landscape. You've proven that curiosity, persistence, and practical application can overcome any initial barriers to understanding and using advanced technology.

The future belongs to people who can effectively collaborate with AI while maintaining their uniquely human capabilities: creativity, empathy, strategic thinking, and ethical judgment. You've positioned yourself perfectly for that future.

Congratulations on completing your AI learning journey. You're now equipped not just with specific tools and techniques, but with the confidence and framework to continue growing and adapting as technology continues to evolve. The most important lesson isn't about AI at all—it's about your own capacity for learning, adapting, and thriving in an ever-changing world.

Welcome to your AI-enhanced future. You're ready to make the most of it.

[10]

AI AT WORK

CAREERS & BUSINESS OPPORTUNITIES

The future of work isn't coming—it's already here, and it's reshaping everything from how we write emails to how companies discover new medicines. If you've been wondering whether artificial intelligence will replace your job, enhance it, or create entirely new opportunities you haven't considered, you're asking exactly the right questions.

The reality is far more nuanced than the headlines suggest. While some roles are indeed changing dramatically, the biggest winners in this transformation won't necessarily be the programmers or data scientists. They'll be the people who learn to work *alongside* AI tools, leveraging them to become more creative, efficient, and valuable in their fields.

Let's explore how this shift is playing out across industries and what it means for your career—whether you're just starting out, looking to pivot, or trying to future-proof the work you already love.

HOW INDUSTRIES ARE ADOPTING AI

Healthcare: Beyond the Stethoscope

Healthcare might seem like the last place you'd expect rapid AI adoption, given its emphasis on human connection and careful regulation. Yet it's becoming one of the most transformed sectors, and not just in ways that grab headlines.

Radiology has been an early adopter, with AI systems now capable of spotting certain cancers in mammograms and CT scans faster and sometimes more accurately than human radiologists. But rather than replacing doctors, these tools are freeing them up to spend more time with patients and focus on complex cases that require human judgment.

Drug discovery represents perhaps the most dramatic change. Traditional pharmaceutical development takes 10-15 years and costs billions of dollars. AI is compressing timelines by identifying promising compounds faster and predicting which ones are likely to fail before expensive clinical trials begin. Companies like Atomwise and Benevolent AI are already bringing AI-discovered drugs to market.

The opportunities here extend far beyond medical professionals. Healthcare now needs people who can bridge the gap between technology and patient care—roles like clinical AI specialists who help hospitals implement new tools, or medical AI trainers who teach staff how to interpret AI-generated insights.

Finance: Numbers Game Revolution

Wall Street was among the first to embrace AI, but the changes go far deeper than algorithmic trading. Fraud detection systems now analyze millions of transactions in real-time, catching suspicious patterns that would take human analysts weeks to identify.

Credit scoring has been revolutionized by machine learning models that consider hundreds of data points instead of just credit history. This has opened up lending opportunities for people who were previously considered "unbankable" while reducing defaults for lenders.

But perhaps most interesting is how AI is democratizing financial advice. Robo-advisors like Betterment and Wealthfront now provide portfolio management services that were once available only to wealthy clients. This isn't just changing how we invest—it's creating new career paths for financial educators who help people understand AI-generated recommendations and fintech product specialists who design user-friendly interfaces for complex financial tools.

Retail and E-commerce: The Personal Touch at Scale

Amazon's recommendation engine was just the beginning. Modern retail AI can predict what you'll want to buy before you know it yourself, optimize pricing in real-time based on demand and competition, and even generate product descriptions automatically.

Inventory management has become incredibly sophisticated. Walmart uses AI to predict demand for 500 million item-store combinations, reducing waste and ensuring products are available when customers want them. Small businesses can now access similar predictive power through tools like Inventory Planner or TradeGecko.

The fashion industry showcases some of the most creative AI applications. Companies like Stitch Fix use machine learning to curate personalized clothing selections, while others employ AI to predict next season's trends by analyzing social media, runway shows, and street style photos.

This transformation is creating entirely new roles: customer experience analysts who interpret AI insights about shopping behavior, demand planners who work with predictive algorithms, and AI product managers who oversee the development of intelligent retail systems.

Manufacturing: Smart Factories Come Alive

Manufacturing might seem like an unlikely candidate for AI transformation, but it's happening faster than almost anywhere else. Predictive maintenance systems monitor equipment continuously, predicting failures before they happen and scheduling repairs during planned downtime rather than after costly breakdowns.

Quality control has been revolutionized by computer vision systems that can spot defects invisible to the human eye. In automotive manufacturing, AI cameras inspect welds, paint jobs, and assembly quality at speeds impossible for human inspectors.

Perhaps most impressively, AI is enabling mass customization—producing unique products at scale without the traditional cost penalties. Nike's Nike By You program uses AI to optimize manufacturing processes for custom shoes, while furniture companies like Nur use AI to create bespoke pieces based on customer preferences and room dimensions.

The career opportunities here extend to industrial AI specialists, smart manufacturing engineers, and human-AI collaboration coordinators who ensure that automated systems work seamlessly with human workers.

Creative Industries: Augmented Imagination

The creative sector has experienced some of the most visible—and controversial—AI developments. Content creation tools like GPT models can write articles, scripts, and marketing copy, while image generators like DALL-E and Midjourney create artwork from text descriptions.

But the most successful applications aren't replacing human creativity—they're amplifying it. Filmmakers use AI for pre-visualization and special effects. Musicians employ AI tools for composition inspiration and audio mastering. Designers leverage AI for rapid prototyping and trend analysis.

Netflix uses AI not just for recommendations but for creating personalized movie thumbnails, choosing different images for the same film based on what's most likely to appeal to each viewer. Marketing agencies use AI to generate multiple ad variations and test them automatically.

New career paths emerging here include AI prompt engineers who specialize in getting the best results from generative AI tools, human-AI creative directors who orchestrate collaborations between human creatives and AI systems, and AI ethics consultants who help creative companies navigate the complex questions around AI-generated content.

FUTURE-PROOFING YOUR CAREER

The Skills That Matter Most

The good news about AI-driven career changes is that the most valuable skills aren't technical ones that require years of study. They're human skills that become more important as AI handles routine tasks.

Critical thinking tops the list. As AI generates more content and analysis, the ability to evaluate, contextualize, and make decisions based on AI outputs becomes crucial. This isn't about learning to code—it's about developing judgment.

Creative problem-solving becomes increasingly valuable as AI handles standard solutions. The problems that remain for humans are the complex, ambiguous ones that require intuition, empathy, and out-of-the-box thinking.

Communication skills are more important than ever. As AI creates more content, the ability to communicate authentically, persuasively, and with genuine human connection becomes a competitive advantage. This includes translating between technical AI outputs and practical business needs.

Adaptability and learning agility matter more than specific technical knowledge, which becomes obsolete quickly. The ability to learn new tools, understand their capabilities and limitations, and integrate them into your workflow is what separates thriving professionals from those struggling to keep up.

Strategic Career Positioning

Become an AI-human bridge. The most valuable professionals will be those who can work effectively with AI tools while maintaining the human elements that AI cannot replicate. This might mean becoming a marketer who uses AI for data analysis but provides human insight into customer psychology, or a teacher who uses AI to personalize learning while providing mentorship and emotional support.

Specialize in AI tool mastery within your field. Every industry will have specific AI applications. Becoming the person who really understands how to get the best results from these tools—and can teach others—positions you as indispensable. This doesn't require a computer science degree, just curiosity and willingness to experiment.

Focus on high-touch, relationship-based work. As AI handles more transactional tasks, human relationships become more valuable. This might mean shifting from data analysis to client consultation, or from content creation to content strategy and relationship building.

Develop meta-skills around AI. Learn to prompt AI tools effectively, evaluate AI outputs critically, and understand the ethical implications of AI decisions. These skills will transfer across tools and industries as AI capabilities evolve.

Building Your AI Toolkit

Start with general-purpose AI tools that apply across industries. ChatGPT for writing and brainstorming, Claude for analysis and research, and tools like Notion AI for productivity. Spend time understanding their strengths and limitations rather than just following tutorials.

Industry-specific tools should be your next focus. If you're in marketing, explore Jasper, Copy.ai, or Persado. In design, try Figma's AI features or Adobe's Sensei. For business analysis, look into platforms like DataRobot or H2O.ai.

Most importantly, practice prompt engineering—the art of communicating effectively with AI systems. This skill is becoming as fundamental as email proficiency was in the 1990s. Good prompts get dramatically better results than poor ones, and this skill translates across all AI tools.

The Portfolio Career Approach

AI is enabling a new kind of career flexibility. As AI tools make it easier to produce professional-quality work across different domains, portfolio careers become more viable. You might combine freelance consulting, course creation, and product development in ways that weren't possible before.

AI tools can help you scale personal expertise. A consultant can use AI to create educational content, a designer can develop digital products, a writer can build software tools—all without hiring large teams or learning to code from scratch.

The key is thinking of AI as your force multiplier rather than a replacement. It allows you to take on projects and explore opportunities that would have been too time-intensive or technically complex before.

HOW TO STAY AHEAD AS AI EVOLVES

The Continuous Learning Mindset

The pace of AI development means that specific tools and techniques become obsolete quickly, but underlying principles remain valuable. Focus on learning how to learn about AI rather than mastering any particular system.

Set up information flows that keep you current without overwhelming you. Follow a few key sources like MIT Technology Review, The Batch (Andrew Ng's newsletter), or industry-specific AI publications. Join professional communities where people share practical experiences with AI tools.

Experiment regularly with new AI tools, even if they're outside your immediate field. The patterns you learn from playing with a new language model or image generator often apply to business tools that arrive later.

Document your experiments. Keep notes on what works, what doesn't, and why. This becomes valuable knowledge that you can share with colleagues or turn into content that establishes your expertise.

Building Your AI Network

The AI space moves too fast for any individual to track all developments. Connect with others who are exploring AI in their fields. This might be through professional associations, LinkedIn groups, local meetups, or online communities.

Share your own experiments and insights. Write about your experiences with AI tools, speak at industry events, or create content about AI applications in your field. Teaching others is one of the best ways to solidify your own understanding and build your reputation as someone who "gets" AI.

Cross-pollinate across industries. Some of the most innovative AI applications come from adapting techniques from one field to another. A marketing professional might learn valuable lessons from how AI is used in healthcare, or a teacher might adapt manufacturing AI concepts for educational assessment.

Ethical Leadership in AI Adoption

As AI becomes more prevalent, ethical considerations become increasingly important career differentiators. Understanding bias in AI systems, privacy implications of data collection, and the social impact of automation decisions will become valuable expertise.

Develop judgment about AI limitations. Know when AI tools are likely to produce unreliable results, understand their biases, and be able to communicate these limitations to colleagues and clients. This expertise becomes crucial as organizations rely more heavily on AI systems.

Champion responsible AI use in your organization. This might mean developing guidelines for AI tool use, training colleagues on limitations and best practices, or serving on ethics committees that oversee AI implementations.

Long-term Strategic Thinking

The current wave of AI development is just the beginning. Anticipate second-order effects of AI adoption in your industry. If AI makes content creation easier, how will that change the value of curation and quality assessment? If AI handles customer service, how will that shift the importance of customer experience design?

Think globally. AI development is happening worldwide, and opportunities may emerge in unexpected places. Remote work enabled by AI tools opens up global collaboration possibilities that weren't practical before.

Consider the human element. As AI becomes more capable, the value of authentically human experiences increases. Think about how your career can emphasize the irreplaceable human elements—creativity, empathy, wisdom, and genuine connection.

The transformation happening now isn't just about technology—it's about reimagining what work can be when humans and artificial intelligence collaborate effectively. The professionals who thrive won't be those who compete with AI, but those who learn to dance with it, leveraging its capabilities while contributing uniquely human value.

Your career in the age of AI doesn't have to be about becoming a different person. It's about becoming a more empowered version of yourself, equipped with tools that amplify your existing strengths and open up possibilities that didn't exist just a few years ago. The question isn't whether AI will change your work—it's how you'll shape that change to create the career and impact you want.

The future belongs to those who start experimenting today. Not with the goal of becoming AI experts overnight, but with the curiosity to explore how these remarkable tools can make their work more meaningful, creative, and effective. That journey begins with the next AI tool you decide to try, the next conversation you have about AI's role in your industry, and the next problem you attempt to solve with human intelligence enhanced by artificial capability.

[11]
AI FOR PERSONAL GROWTH & EVERYDAY LIFE

The alarm goes off at 6:30 AM. Instead of groggily checking social media, imagine reaching for your phone and asking your AI assistant to help you plan the perfect morning routine based on your energy levels, the weather, and your goals for the day. By 7:15 AM, you're following a personalized workout routine an AI fitness coach designed specifically for your body type and available equipment. Over breakfast, you're practicing French conversation with an AI tutor who never judges your pronunciation mistakes and adjusts the difficulty level in real-time based on your progress.

This isn't science fiction—it's Tuesday morning for millions of people who've discovered how to weave artificial intelligence into the fabric of their daily lives. While most discussions about AI focus on business applications or world-changing technologies, the most immediate and transformative impact might be happening in the quiet moments of personal growth, learning, and self-improvement.

THE RISE OF AI AS YOUR PERSONAL GROWTH PARTNER

Traditional personal development often came with barriers: expensive coaches, rigid class schedules, one-size-fits-all programs, or the intimidation factor of joining group activities. AI dissolves these obstacles by offering something unprecedented—a completely personalized, infinitely patient, and constantly available partner in your journey of self-improvement.

Unlike human coaches or teachers who have limited time and energy, AI systems can provide round-the-clock support, instantly adapt to your learning style, and track your progress with microscopic precision. More importantly, they create a judgment-free zone where you can explore new interests, ask "stupid" questions, and fail spectacularly without embarrassment.

The shift is profound. Instead of trying to fit yourself into existing programs or methodologies, you can now design growth experiences that work specifically for your personality, schedule, lifestyle, and goals. AI becomes the ultimate collaborative partner—one that brings vast knowledge and analytical capabilities while letting you maintain complete control over your direction and pace.

AI AS YOUR PERSONAL ASSISTANT: BEYOND SCHEDULING AND REMINDERS

When most people think of AI assistants, they imagine setting timers or asking about the weather. The reality is far more sophisticated. Modern AI assistants can serve as comprehensive life management partners, helping you optimize everything from your daily routines to long-term goal achievement.

Consider habit formation. Traditional advice tells you to pick one habit and stick with it for 30 days. An AI assistant can track multiple habit attempts simultaneously, identify patterns in your success and failure rates, and dynamically adjust your approach. It might notice that you're 40% more likely to exercise when it's sunny, 60% more consistent with morning routines when you go to bed before 10 PM, and completely unsuccessful with meditation when you attempt it right after meals.

Armed with this data, your AI assistant doesn't just remind you to meditate—it suggests the optimal time based on your personal success patterns, offers alternative techniques when your usual approach isn't working, and even adjusts your entire day's schedule to create conditions that support your goals.

Decision-making becomes another area where AI assistance shines. Instead of agonizing over choices large and small, you can engage your AI in structured decision-making processes. Planning a career change? Your AI can help you map out pros and cons, research market trends, identify skill gaps, and even simulate potential outcomes based on different choices. Choosing between vacation destinations? It can analyze your past travel preferences, current interests, budget constraints, and even seasonal factors to provide genuinely personalized recommendations.

The key difference from traditional decision-making tools is the conversational nature of modern AI. You're not filling out forms or answering multiple-choice questions—you're having a dialogue with a system that can understand context, ask clarifying questions, and adapt its approach based on your thinking style.

AI AS YOUR TUTOR: PERSONALIZED LEARNING REVOLUTION

The educational applications of AI represent perhaps the most exciting frontier for personal growth. Unlike traditional learning environments where one teacher manages dozens of students, AI tutors can provide truly individualized instruction that adapts in real-time to your comprehension, interests, and learning preferences.

Language learning showcases this potential beautifully. While apps like Duolingo introduced gamification to language acquisition, AI-powered conversation partners take it several steps further. These systems can engage in natural dialogue at your exact proficiency level, gradually introducing new vocabulary and grammar concepts within context rather than through isolated drills.

More sophisticated AI tutors can even adapt their teaching personality to match your preferences. Some learners respond well to encouraging, cheerleader-type feedback. Others prefer direct, no-nonsense correction of mistakes. Still others learn best through humor and playful interaction. AI systems can experiment with different approaches and settle on the style that produces the best results for your specific psychology.

Subject mastery becomes more accessible across virtually any field. Want to understand quantum physics? An AI tutor can start with your current level of scientific background and build understanding step by step, using analogies and examples that resonate with your personal interests and experiences. Struggling with photography composition? An AI can analyze your current photos, identify specific areas for improvement, and provide targeted exercises that address your particular weaknesses.

The Socratic method—learning through guided questioning rather than direct instruction—becomes particularly powerful when implemented by AI. Instead of simply providing answers, sophisticated AI tutors can guide you to discover insights through carefully crafted questions that build on your existing knowledge and push your thinking in new directions.

AI AS YOUR COACH: MOTIVATION, ACCOUNTABILITY, AND GROWTH

Coaching represents one of the most nuanced applications of AI in personal development. Effective human coaches combine technical knowledge, emotional intelligence, motivational skills, and the ability to provide both support and accountability. While AI may not replicate the full depth of human coaching relationships, it offers unique advantages that complement traditional coaching approaches.

Consistency stands out as AI's greatest coaching strength. Human coaches have bad days, personal challenges, and limited availability. AI coaches show up every day with the same level of attention and energy. This consistency becomes particularly valuable for maintaining momentum during the inevitable ups and downs of any growth journey.

Pattern recognition allows AI coaches to identify trends and correlations that might escape human observation. Your AI coach might notice that your productivity drops every third Thursday, that you're most creative between 2-4 PM on days when you exercise in the morning, or that your mood correlates strongly with how much water you drink. These insights can inform more effective strategies and interventions.

Personalized motivation becomes possible when AI systems learn what drives you specifically. Some people respond to achievement-based motivation, others to purpose-driven messaging, still others to social pressure or friendly competition. AI coaches can experiment with different motivational approaches and focus on the techniques that actually move the needle for your personality type.

The **24/7 availability** factor creates opportunities for real-time coaching interventions. Feeling unmotivated at 10 PM on a Sunday? Your AI coach is available for a pep talk or strategy session.

Experiencing a breakthrough moment during your morning jog? You can immediately share it and get feedback that builds on the momentum rather than waiting for your next scheduled coaching session.

USING AI FOR HOBBIES AND CREATIVE EXPLORATION

One of the most delightful applications of AI in personal growth involves exploring new hobbies and creative pursuits. The traditional barrier to trying new activities—not knowing where to start, lacking basic knowledge, or feeling intimidated by more experienced practitioners—disappears when you have an AI companion ready to guide your exploration.

Creative writing offers a perfect example. Instead of staring at a blank page, you can collaborate with AI to generate story ideas, develop characters, plot out narratives, or even work through writer's block. The AI becomes a creative sparring partner—someone to bounce ideas off, get unstuck with, or explore "what if" scenarios alongside.

Some writers worry that AI assistance might diminish the authenticity of their creative work. In practice, most discover that AI serves more as a creative catalyst than a replacement for human imagination. The AI might suggest an unexpected plot twist that sparks your creativity in a completely different direction, or help you explore character motivations you hadn't considered. The final creative decisions and the voice behind the work remain distinctly yours.

Music creation has been democratized through AI tools that can generate melodies, suggest chord progressions, or even create backing tracks in any style you specify. You don't need years of music theory education to start experimenting with composition. The AI handles technical complexity while you focus on the creative vision and emotional expression.

Visual arts benefit similarly from AI partnerships. Whether you're interested in digital painting, photography composition, or even traditional mediums, AI can provide technique suggestions, composition feedback, and creative prompts that push your artistic exploration in new directions. Some artists use AI-generated images as starting points for traditional paintings, while others incorporate AI-created elements into mixed-media works.

Learning instruments becomes less intimidating when AI can provide patient, personalized instruction that adapts to your progress rate and musical interests. Unlike human teachers who might have strong opinions about proper technique or musical styles, AI instructors can accommodate any genre preference and work within your natural learning patterns.

SELF-IMPROVEMENT THROUGH AI-GUIDED REFLECTION

Personal growth often requires stepping outside our own perspectives to gain clarity about patterns, blind spots, and potential directions for development. AI excels at facilitating this kind of structured self-reflection through guided questions, pattern analysis, and objective feedback.

Journaling becomes more powerful when enhanced by AI analysis. Instead of simply recording daily events, you can engage in dialogue with an AI that asks probing questions, identifies recurring themes in your entries, and helps you connect dots between different areas of your life. The AI might notice that your energy levels correlate with certain types of social interactions, or that your most productive weeks follow specific patterns of rest and activity.

Goal setting and tracking benefit from AI's ability to break down abstract aspirations into concrete, actionable steps. Want to "become more confident"? An AI can help you define what confidence looks like in your specific context, identify situations where you already demonstrate confidence, and design experiments that gradually expand your comfort zone.

The **objective perspective** that AI provides can be invaluable for personal development. Friends and family often have their own biases and emotional investments in your choices. AI can offer feedback and observations without hidden agendas, personal relationships to navigate, or emotional reactions to manage.

Skill gap analysis becomes systematic when AI can compare your current capabilities with your desired goals, then map out specific learning paths and milestones. Instead of vague advice to "work on communication skills," you might receive specific suggestions like "practice active listening techniques in low-stakes conversations this week, then gradually apply them in more challenging situations."

DESIGNING YOUR AI-ENHANCED LIFESTYLE: A PRACTICAL FRAMEWORK

The goal isn't to replace human connection and personal agency with AI systems, but rather to create a symbiotic relationship where artificial intelligence amplifies your natural capabilities and removes friction from your growth journey. Designing an AI-enhanced lifestyle requires thoughtful consideration of where AI can add genuine value without creating dependency or diminishing authentic human experiences.

Morning Optimization

Start by examining your morning routine through an AI lens. Instead of checking social media immediately upon waking, consider beginning each day with a brief AI conversation that helps you set intentions, review priorities, and optimize your schedule based on energy levels and commitments.

Your AI might analyze sleep data from wearable devices and suggest modifications to your morning routine based on sleep quality. Poor sleep night? Maybe the AI recommends a gentler workout and extra time for mindfulness. Great sleep? Perhaps it suggests tackling your most challenging creative work first thing.

Personalized news and information consumption becomes more intentional when AI curates content based on your genuine interests and growth goals rather than engagement algorithms designed to capture attention. Instead of random social media scrolling, you might receive a customized briefing that includes industry trends relevant to your career, learning resources for your current skill development goals, and inspiring content that aligns with your values.

Learning and Skill Development Integration

Rather than treating learning as a separate activity confined to specific times or locations, AI enables you to weave continuous growth into your daily routine. Microlearning becomes possible during previously "dead" time—commutes, waiting periods, exercise sessions, or routine tasks.

Your AI tutor might provide 5-minute language lessons tailored for your walk to work, offer photography tips while you're actually taking photos, or share business insights relevant to a project you're currently working on. The key is contextual relevance—learning that connects directly to your immediate experiences and goals.

Skill combination becomes more strategic when AI can identify connections between different areas of development. Learning graphic design? Your AI might suggest combining it with your interest in psychology by exploring the psychological principles behind visual communication. Developing public speaking skills?

The AI could connect this with your hobby of storytelling to create more engaging presentation techniques.

Social and Relationship Enhancement

AI can enhance rather than replace human connections by helping you become more intentional and effective in your social interactions. Conversation preparation for important meetings, social events, or difficult conversations becomes more thoughtful when AI can help you anticipate different scenarios, practice key messages, and consider multiple perspectives.

Conflict resolution skills can be developed through AI-guided role-playing and scenario analysis. Instead of entering challenging conversations unprepared, you can explore different approaches, practice de-escalation techniques, and develop strategies that align with your communication style and values.

Relationship maintenance becomes more systematic when AI helps you stay connected with important people in your life. Rather than letting friendships drift due to busy schedules, your AI assistant might suggest thoughtful ways to reach out, remind you of important events in friends' lives, or help you plan meaningful shared experiences.

Creative Project Management

Long-term creative projects often stall due to lack of structure, motivation challenges, or difficulty breaking down complex goals into manageable tasks. AI project management can transform ambitious creative visions into step-by-step reality.

Progress tracking becomes more nuanced when AI can identify both quantitative progress (words written, hours practiced, skills learned) and qualitative improvements (style development, creative risk-taking, emotional expression). This comprehensive view helps maintain motivation during inevitable periods when progress feels slow.

Creative problem-solving benefits from AI's ability to suggest unexpected connections, alternative approaches, and fresh perspectives when you feel stuck. The AI becomes a **brainstorming partner** that can generate ideas at any time, helping you push through creative blocks and explore new directions.

Health and Wellness Integration

Physical and mental wellness form the foundation for all other growth activities. AI can help optimize your wellness routine by identifying patterns, suggesting improvements, and providing consistent support for healthy habits.

Nutrition optimization becomes personalized when AI analyzes your energy levels, mood, performance, and health goals in relation to your eating patterns. Instead of following generic diet advice, you can develop eating strategies that work specifically for your body, lifestyle, and preferences.

Exercise personalization extends beyond generic workout programs to consider your current fitness level, available time, equipment access, injury history, and motivation patterns. Your AI fitness coach can adjust workouts in real-time based on how you're feeling, what you have time for, and what equipment is available.

Stress management becomes proactive when AI can recognize early warning signs of burnout or overwhelm and suggest interventions before problems become serious. The AI might notice patterns in your sleep, mood, productivity, and social interactions that indicate rising stress levels, then recommend specific strategies that have worked for you in the past.

BUILDING YOUR AI-ENHANCED LIFE: A STEP-BY-STEP MINI-PROJECT

Let's move from theory to practice by designing your personalized AI-enhanced lifestyle. This isn't about adopting someone else's system—it's about creating an approach that amplifies your unique strengths, addresses your specific challenges, and supports your individual goals.

Step 1: Life Audit and Goal Clarification

Begin by conducting an honest assessment of your current situation and desired improvements. Instead of vague aspirations like "be healthier" or "learn new skills," work with AI to develop specific, measurable goals that align with your values and circumstances.

Start a conversation with your chosen AI assistant about your life satisfaction across different areas: career development, health and fitness, relationships, creative expression, learning and growth, and personal fulfillment. Ask the AI to help you rate each area and identify the top three priorities for improvement.

For each priority area, engage the AI in a goal refinement process. If you want to improve your health, AI can help you define what "better health" means specifically for you—is it increased energy, weight management, better sleep, stress reduction, or physical fitness? Once you've clarified your goals, the AI can help you establish baseline measurements and define success metrics.

Step 2: Daily Routine Optimization

Map out your current daily routine in detail, then work with AI to identify optimization opportunities. The goal isn't to create a rigid schedule, but rather to build in systems and supports that make positive choices easier and more automatic.

Consider your energy patterns throughout the day. Most people have natural peaks and valleys in alertness, creativity, and motivation. AI can help you identify these patterns through simple tracking and then suggest how to align different types of activities with your natural rhythms.

Design AI touchpoints throughout your day, specific times when you'll engage with AI systems to support your goals. This might include morning intention-setting, midday energy check-ins, learning sessions during commute time, evening reflection, and weekly progress reviews.

Step 3: Learning and Skill Development Strategy

Identify the skills and knowledge areas that would most impact your goals and life satisfaction. Rather than trying to learn everything at once, prioritize based on immediate applicability, personal interest, and long-term value.

Work with AI to create a personalized curriculum for each priority learning area. This should include specific resources, practice opportunities, progress milestones, and ways to apply new knowledge in your daily life. The AI can help you balance different learning modalities—reading, watching, listening, practicing, teaching others—to match your preferred learning style while ensuring comprehensive skill development.

Create learning integration points where new knowledge connects with your existing routines and responsibilities. Instead of treating learning as an additional burden on your schedule, find ways to weave skill development into activities you're already doing. Learning Spanish? Practice during your evening walk. Developing leadership skills? Apply new techniques in your current work projects. Exploring photography? Use your daily commute as an opportunity to practice composition and lighting observation.

Step 4: Implementation and Iteration

Start small with pilot experiments rather than attempting to transform your entire lifestyle overnight. Choose one or two areas where AI enhancement can make an immediate positive impact, implement these changes for two weeks, then evaluate and adjust based on results.

Track both quantitative and qualitative measures of progress. Numbers tell part of the story, hours spent learning, workout consistency, productivity metrics, but pay equal attention to subjective improvements in mood, energy, satisfaction, and confidence. Your AI can help you identify patterns and correlations that might not be immediately obvious.

Build in regular review and adjustment cycles. What works in winter might not work in summer. What motivates you during high-energy periods might need modification during stressful times. The beauty of AI-enhanced living is its adaptability, your system can evolve as you do.

THE PATH FORWARD

The integration of AI into personal growth represents more than technological adoption. It's an opportunity to become more intentional about how you develop, learn, and evolve as a person. The goal isn't to optimize yourself like a machine but to remove friction from the pursuits that matter most to you and create more opportunities for authentic growth and self-expression.

As you begin experimenting with AI-enhanced living, remember that the technology serves you, not the other way around. The most successful implementations feel natural and supportive rather than demanding or overwhelming. You remain the author of your own growth story. AI simply provides better tools for writing it.

Start where you are, with what you have, and let curiosity guide your exploration. The future of personal development isn't about choosing between human wisdom and artificial intelligence, but about combining both in service of becoming the person you want to be. Your AI-enhanced life begins with the next conversation, the next question, the next small step toward who you're becoming.

[12]
FROM BEGINNER TO CONFIDENT CREATOR

You've made it this far. You've experimented with ChatGPT, maybe dabbled with image generators, and probably felt that exciting rush when AI helped you solve a problem that would have taken hours before. But now what? How do you go from someone who knows a few AI tricks to someone who confidently integrates these tools into their daily workflow?

This chapter isn't about learning more AI tools. It's about becoming the kind of person who uses AI naturally and effectively. Avoiding common traps, building sustainable habits, and knowing where to turn when you need to level up are the three key differences between someone who plays with AI occasionally and someone who consistently leverages it.

Let's tackle each of these head-on.

THE HIDDEN TRAPS THAT TRIP UP NEW AI USERS

The Shiny Object Syndrome

Every week, there's a new AI tool making headlines. One day it's a revolutionary video generator, the next it's an AI that can code better than most programmers. The temptation to try everything is overwhelming, especially when social media influencers make each new tool sound like the missing piece of your productivity puzzle.

Here's what actually happens: You sign up for five new AI platforms in a month. You spend twenty minutes with each one, get confused by the interface, don't see immediate results, and move on to the next shiny thing. Meanwhile, the tools you already know—the ones that could genuinely transform your work, sit unused because you're always chasing the latest release.

The fix? Pick your core tools and stick with them until they become second nature. If ChatGPT is working for your writing, don't abandon it for the newest text generator just because it has a sleeker interface. Master what you have before you expand your toolkit.

The Perfectionist's Paradox

Many newcomers approach AI with the expectation that their first prompt should produce perfect results. When it doesn't, they assume they're doing something wrong or that AI isn't as powerful as advertised. This leads to either giving up entirely or spending hours crafting the "perfect" prompt before even trying.

The reality is messier and more forgiving. AI works best through iteration. Your first attempt should be good enough to get started, then you refine from there. Professional prompt engineers don't create masterpieces on their first try. They build them through multiple rounds of testing and adjustment.

The mindset shift: Think of your first prompt as a rough draft, not a final product. Give AI something to work with, see what comes back, then guide it toward what you actually need.

The Context Collapse

This one catches almost everyone. You're having a productive conversation with ChatGPT, getting great results, when suddenly the responses start feeling off. The AI seems to have forgotten what you were working on, or it's giving you generic advice instead of the specific help you need.

What happened? You hit the context limit. AI models can only "remember" a certain amount of conversation history. When you exceed that limit, the early parts of your conversation drop out of memory, and the AI loses track of your specific situation.

The solution: Start fresh conversations for new projects. If you're switching from writing emails to planning a marketing campaign, open a new chat. Keep your conversations focused on single topics or related tasks.

The Delegation Trap

As AI gets better, it's tempting to hand over more and more responsibility. "Why should I learn to write better when AI can write for me?" This thinking leads to a dangerous dependency where your own skills atrophy while you become increasingly reliant on AI output.

The most effective AI users don't replace their abilities—they amplify them. They use AI to generate first drafts, then apply their judgment to edit and improve. They let AI handle research, then use their expertise to synthesize insights. They maintain and develop their core competencies while using AI to eliminate busy work.

BUILDING CONSISTENCY WITHOUT BURNING OUT

The Daily Practice Approach

Consistency doesn't mean using AI for everything, every day. It means integrating AI into your existing routines in sustainable ways. Instead of dedicating an hour each evening to "AI practice," look for natural integration points in your current workflow.

Maybe you already write a daily task list—try having ChatGPT help you prioritize it. If you read industry news each morning, ask Claude to summarize the key points from articles you don't have time to read fully. The goal is to make AI feel like a natural extension of what you already do, not an additional burden.

The Progressive Challenge System

Once you're comfortable with basic AI interactions, you need a way to gradually expand your capabilities without overwhelming yourself. Here's a framework that works:

Week 1-2: Focus on one specific use case. If you're using AI for writing, stick to one type of content—maybe email drafts or social media posts.

Week 3-4: Add complexity to your existing use case. Instead of just asking for email drafts, start requesting specific tones, lengths, or styles.

Week 5-6: Introduce a second use case in a completely different area. If you started with writing, try data analysis or creative brainstorming.

Week 7-8: Look for ways to chain your use cases together. Use AI to research a topic, then to write about it, then to create a presentation based on your writing.

This progression keeps you challenged without creating cognitive overload. Each step builds on the previous one, creating a natural learning curve.

The Energy Management Strategy

Not all AI tasks require the same mental energy. Learning to match your AI activities to your energy levels prevents burnout and maintains enthusiasm.

High Energy (Morning/Peak Hours): Complex problem-solving, learning new tools, creative projects that require significant input and iteration.

Medium Energy (Afternoon): Routine improvements, content editing, data organization tasks where AI does the heavy lifting but you provide direction.

Low Energy (Evening/Off Days): Simple automation, content consumption (having AI summarize articles or videos), maintenance tasks like organizing your AI prompt library.

By aligning your AI activities with your natural energy rhythms, you'll maintain consistency without feeling drained.

YOUR CURATED LEARNING PATH FORWARD

Essential Tools for Different Needs

Rather than listing every AI tool available, here are the proven platforms organized by what you're trying to accomplish:

For Content Creation:

- **ChatGPT or Claude** for writing assistance, editing, and brainstorming

- **Grammarly** for polishing and tone adjustment
- **Canva's AI features** for design elements and layouts

For Business and Productivity:

- **Notion AI** for note organization and project planning
- **Zapier** for connecting your AI tools to other apps
- **Calendly's scheduling assistant** for meeting coordination

For Learning and Research:

- **Perplexity** for research with source citations
- **YouTube's transcript + AI summary tools** for video content consumption
- **Readwise** combined with AI summarization for article processing

For Creative Projects:

- **Midjourney** or **DALL-E** for image generation
- **RunwayML** for video creation and editing
- **ElevenLabs** for voice synthesis and audio content

The key is to start with one tool from one category, master it completely, then gradually expand to other areas based on your actual needs, not what sounds interesting in theory.

Communities That Actually Help

The AI space is full of communities, but many focus on technical discussions that overwhelm beginners or promotional content that adds more noise than value. Here are communities known for practical, beginner-friendly support:

Reddit Communities:

- **r/ChatGPT** for practical tips and real-world use cases
- **r/artificial** for broader AI discussions without too much technical depth
- **r/productivity** where AI tools are discussed in the context of getting things done

Discord Servers:

- Many AI tool companies run their own Discord servers where users share tips and troubleshoot issues. These are often more helpful than general AI communities because the focus is narrow and practical.

Professional Networks:

- **LinkedIn AI groups** tend to focus on business applications and career development
- **Facebook groups** organized around specific professions (AI for marketers, AI for writers, etc.)

Learning Resources That Respect Your Time

For Quick Updates (5-10 minutes daily):

- **AI newsletters** like "The Rundown AI" or "AI Breakfast" that summarize developments without hype
- **Twitter/X lists** following practical AI users rather than researchers or venture capitalists
- **YouTube channels** focused on tutorials rather than news (channels like "AI Advantage" or "All About AI")

For Deeper Learning (Weekly 30-60 minute sessions):

- **Coursera's AI for Everyone** course by Andrew Ng—technical enough to be credible, accessible enough for non-programmers
- **Khan Academy's AI courses** for foundational concepts explained clearly
- **MasterClass offerings** on AI topics, which tend to be well-produced and practical

For Problem-Solving (As needed):

- **Stack Overflow** has an increasing number of AI-related questions and answers
- **GitHub repositories** with prompt libraries and examples
- **Company documentation** for the specific tools you use regularly

The Monthly Check-In System

To maintain momentum without getting overwhelmed, establish a monthly review process:

Week 1: Assess what's working. Which AI tools or techniques have you actually used? Which ones are gathering digital dust?

Week 2: Identify gaps. What problems are you still solving manually that AI could handle? What new challenges have emerged in your work that AI might address?

Week 3: Experiment with one new approach. This could be a new tool, a different way of using an existing tool, or a completely different application area.

Week 4: Document and organize. Update your prompt library, bookmark useful resources, and note what you want to explore next month.

This system prevents both stagnation and overwhelm by creating regular touchpoints without demanding constant attention.

MAKING AI WORK FOR YOUR SPECIFIC SITUATION

The Personal AI Strategy Canvas

Instead of trying to use AI like everyone else, develop an approach that fits your unique circumstances. Consider these factors:

Time Availability: If you have 10 minutes daily, focus on quick wins like email drafting or task prioritization. If you have an hour weekly, tackle larger projects like content planning or competitive research.

Learning Style: Visual learners benefit from AI-generated diagrams and charts. Auditory learners can use AI to convert written content to audio formats. Kinesthetic learners should focus on AI tools that enhance hands-on activities.

Work Context: Remote workers might emphasize AI for communication and collaboration. In-person teams might focus on AI for presentation and planning. Solopreneurs often get the most value from AI that handles administrative tasks.

Risk Tolerance: Conservative users should start with well-established tools and clear use cases. Early adopters can experiment with beta features and newer platforms.

Troubleshooting Common Sticking Points

When AI outputs feel generic: You're not providing enough context. Include background information, specify your audience, mention your goals, and give examples of what you like and dislike.

When you can't think of what to ask: Keep a running list of frustrating or time-consuming tasks. These are your best candidates for AI assistance. Don't wait for inspiration—document problems as they occur.

When results are inconsistent: You're probably using different prompts for similar tasks. Develop templates for your common use cases and refine them over time rather than starting from scratch each time.

When you feel like you're not improving: You're likely not challenging yourself enough. If AI feels easy, you're probably underutilizing it. Try more complex tasks or combine multiple AI capabilities in single projects.

YOUR NEXT STEPS

You're no longer a beginner. You have the foundational knowledge and practical experience to use AI effectively. The question now is how to build on that foundation systematically.

Start with an honest assessment: What three AI applications have provided the most value in your current workflow? Double down on those. Get so comfortable with them that using AI feels as natural as using a calculator or spell-check.

Then, identify one area where you're still struggling with manual processes. Maybe it's research for projects, maybe it's creating presentations, maybe it's managing your schedule. Pick one area and commit to finding an AI solution over the next month.

Finally, connect with others who are on a similar journey. AI is evolving too quickly for anyone to master alone. The most successful AI users are those who learn from and share with a community of practice.

The goal isn't to become an AI expert—it's to become someone who uses AI so naturally and effectively that it fades into the background of your enhanced capabilities. You're not trying to keep up with technology; you're making technology keep up with your ambitions.

GUIDE A CURIOUS MIND INTO THE WORLD OF AI

One of life's quiet truths is that sharing what we've learned can shorten the path for someone else.

Have you ever wished that, when you were starting something new, someone had been there to say, *"It's not as hard as it looks—you've got this"*?

Imagine a beginner staring at the buzz around **AI, Deep Learning, and ChatGPT**, but feeling overwhelmed and convinced it's "too technical" for them. They're curious, but hesitant—wondering if they're capable of understanding, let alone creating something meaningful with AI. Your reflections on *Generative AI for Beginners: Escape the Learning Curve! A 7-Day Guide to Learn AI, Deep Learning, and ChatGPT — Master Artificial Intelligence and Build Creative Projects Even If You're Not Tech-Savvy* could be the encouragement they need to take that leap.

That's why I'd be deeply grateful if you could leave an **honest review**. Your words aren't just feedback—they're a beacon for the next learner who feels uncertain but hopeful.

By sharing your thoughts, you:

- **Empower future readers** to see AI as approachable and within their reach.
- **Support the mission** of making advanced ideas simple, practical, and fun.
- **Help demystify AI**, reminding beginners that they don't need to be "tech-savvy" to start.

Thank you, truly, for lending your voice to this journey. And if you know someone who's curious about AI but doesn't know where to begin, I'd be honored if you passed this book along to them, too.

CONCLUSION

Remember where you started this journey? Maybe you were one of those people who cringed when friends casually mentioned "prompting ChatGPT" or felt a wave of panic when colleagues discussed automating their workflows with AI tools. Perhaps you nodded along in meetings while secretly wondering what everyone was talking about, or maybe you avoided AI conversations altogether because they felt like stepping into a foreign country without a translator.

That person feels distant now, doesn't it?

Look at what you have accomplished. You have demystified the buzzwords that once seemed impenetrable. You understand that machine learning is not magic; it is pattern recognition at scale. You know that neural networks are inspired by how our brains process information, not by some robotic intelligence plotting world domination. You have discovered that natural language processing simply means teaching computers to understand human communication, and that computer vision helps machines interpret images the way we naturally do.

More importantly, you have moved beyond theory into practice. You have learned to craft prompts that get meaningful results, discovered AI tools that genuinely make your work easier, and developed the confidence to experiment with new applications as they appear. You are no longer intimidated by technology; you are empowered by it.

This transformation did not happen overnight, and it did not require you to become a programmer or data scientist. You simply needed someone to explain AI in human terms, give you practical examples you could relate to, and show you how to start small with achievable wins. That is exactly what curious, determined people like you do best: take complex ideas, break them down, and make them useful.

AI AS YOUR CREATIVE AND PROFESSIONAL PARTNER

One of the biggest shifts in your thinking has probably been reframing AI from a threat into a collaborator. The media loves dramatic headlines about AI replacing human jobs, but your own experience has shown you something very different. AI tools do not replace human creativity, judgment, or emotional intelligence; they amplify these uniquely human strengths.

Think about how this plays out in real scenarios. When you use AI to brainstorm marketing ideas, you are not outsourcing creativity; you are expanding it. The AI generates dozens of possibilities quickly, but you still decide which concepts align with your brand, which messages will resonate with your audience, and how to execute the ideas with authenticity and strategic thinking.

When you use AI writing assistants to draft emails or create content, you are not becoming less skilled as a communicator. Instead, you are freeing up mental energy from routine tasks so you can focus on higher-level strategy, relationship building, and creative problem-solving. The AI handles the first draft, but you bring the insight, experience, and personal voice that transform good writing into something memorable.

This partnership model extends to nearly every profession. Small business owners use AI to analyze customer data and identify trends, but they make the strategic decisions about how to respond. Artists and designers use AI image generators as creative starting points, then apply their aesthetic judgment and technical skills to refine the results. Students and professionals use AI research assistants to gather information more efficiently, then analyze it through their own expertise and perspective.

The key insight you have gained is that AI works best when it complements human capabilities rather than trying to replace them. Machines excel at processing large amounts of data, recognizing patterns, and generating multiple options quickly. Humans excel at understanding context, making nuanced decisions, building relationships, and giving meaning to information. When these strengths come together, the results are far greater than what either could achieve alone.

YOUR INVITATION TO KEEP EXPLORING AND CREATING

Your AI education does not end with this book. It is only beginning. You now have the foundational knowledge and practical skills to confidently explore new AI tools as they appear, adapt to technological changes, and continue growing your capabilities.

The landscape will keep evolving, and that is something to look forward to rather than fear. New AI applications will emerge in fields such as healthcare, education, environmental science, and creative industries. Some will transform how we live and work, while others will fade away as trends. With your AI literacy, you can tell the difference between genuine innovation and empty hype, adopting only what truly supports your goals.

Start with curiosity, not perfection. You do not need to master every new AI tool that launches; that would be impossible and unnecessary. Approach new technologies with the same experimental mindset you have developed throughout this journey. Try tools that match your interests or professional needs. Give yourself permission to explore without pressure to be perfect. Some experiments will lead to valuable additions to your toolkit, while others will simply expand your understanding of what is possible.

Build on your foundation strategically. You now understand the difference between narrow AI, which is designed for specific tasks, and the broader AI ecosystem. You can distinguish between real capabilities and marketing buzz. Use this knowledge to choose wisely which tools deserve your time and attention. Focus on those that either solve real problems or open creative possibilities that excite you.

Share your knowledge and learn from others. You are no longer the person who stays silent in AI conversations; you now have valuable insights to share. Help friends and colleagues who still feel uncertain about AI. Show them practical examples and guide them through their first experiments. At the same time, stay connected with other curious learners who explore AI in different fields. Their discoveries will inspire you to keep growing.

THE CONFIDENCE TO SHAPE YOUR AI-ENHANCED FUTURE

The most important outcome of this journey is not just understanding AI. It is developing the confidence to shape how AI fits into your life rather than letting technology happen to you.

You now have the knowledge to make intentional choices about which AI tools to use, how to integrate them into your workflows, and when to rely on human judgment instead. You understand enough about how these systems function to use them effectively and recognize their limitations. You can now discuss AI's role in your industry, community, and personal projects with clarity and confidence.

This confidence goes beyond using today's AI tools. You have developed technological resilience—the ability to adapt as innovations evolve without feeling overwhelmed or left behind. When the next big AI breakthrough makes headlines, you will respond with curiosity instead of anxiety. When colleagues discuss new AI-powered solutions, you will contribute meaningfully rather than feeling excluded.

Most importantly, you understand that being AI-ready does not mean changing who you are or abandoning your existing skills. It means enhancing them, expanding your creative range, and approaching challenges with a broader set of tools. Your human judgment, creativity, empathy, and expertise remain central to everything you do. AI simply strengthens these qualities and helps you express them more effectively.

The future you are stepping into is not one where humans compete with machines, but one where thoughtful people like you collaborate with AI to solve problems, create meaningful work, and build better outcomes for everyone. You are ready for that future because you have taken the time to understand how AI truly works, learned how to use it effectively, and maintained a balanced perspective on both its potential and its limits.

Welcome to your AI-enhanced future. You are not just ready for it. You are prepared to help shape it.

FOR

Made in United States
Troutdale, OR
11/03/2026

56845400R00100